FOLENS SCIENCE

BOOK THREE

CHRIS BUTCHER
WENDY DEWICK

ALAN JONES

FOLENS SCIENCE: BOOK THREE

© 1992 Folens Limited, on behalf of the authors.

First published 1992 by Folens Limited, Dunstable and Dublin.

Folens Limited, Albert House, Apex Business Centre, Boscombe Road, Dunstable LU5 4RL, England.

ISBN 1 85276104-0

Printed in Singapore by Craft Print.

CONTENTS

WHAT SCIENTISTS DO

The people who have written this book are scientists. We have thought about what makes a good scientist and what is involved in doing good science. When you use this book you will be working as a young scientist.

SCIENTISTS WONDER ABOUT THINGS

They wonder why things happen
They wonder how things work
They wonder how things can be improved

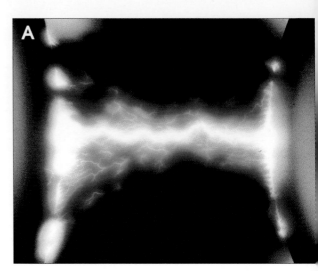

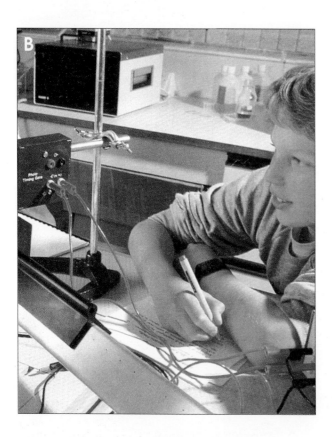

SCIENTISTS RECORD

They write down their observations
They make drawings
They draw tables and graphs
They take photographs
They use computers
They use sound or video tape

SCIENTISTS DO EXPERIMENTS

They choose equipment
They put equipment together
They work safely
They follow instructions
They observe and measure

You are the scientist now. Whenever you do an experiment, indicated by this sign, you should look at these pages. Use them as a checklist.

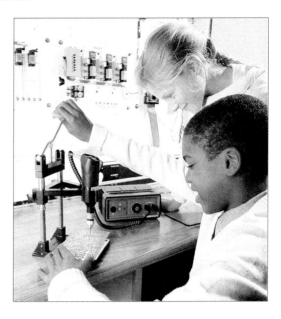

SCIENTISTS PLAN

They decide on the aims of their experiments
They search for ideas
They state hypotheses
They identify variables and control them
They make predictions
They suggest methods of working

SCIENTISTS EVALUATE

They look for patterns
They look for errors
They make inferences
They check predictions
They check if aims were achieved
They look for applications of their work

SCIENTISTS COMMUNICATE

They talk about their experiments
They listen to others
They make posters and models to
show what they have found out

BITS MOVING

What's that smell? Where's it coming from?

Why does this police dog need a good sense of smell?

There are some smells that we like and others that we try to get away from. You can't see the smells - it's the job of your nose to detect the "smell chemicals" as they arrive after travelling through the air. The "smell chemicals" drifting towards you from a plate of food covered with vinegar or strong-smelling sauce must be very small.

Over 2000 years ago, a famous Greek thinker called Democritus imagined that if you took a piece of metal foil and cut it in half and then half again, and so on, using amazingly small scissors, eventually you would come to the smallest piece of metal possible (**B**).

Democritus called this smallest possible piece an **ATOM** (meaning *"not cuttable"* in Greek).

How many times can Democritus cut the foil in half?

The most likely picture we have these days agrees with the basic idea that Democritus produced. Modern scientists have taken his idea much further and designed experiments to find out if their predictions are correct.

The best evidence suggests that the very small atoms do not usually stay apart from each other but join up to make larger pieces called **MOLECULES**. The smell you notice is likely to be made up of small molecules of the smell chemical carried in the air (**C**).

How do "smell chemicals" reach your nose?

Teacher demonstration

Your teacher will place a bottle of a smelly chemical in the corner or the middle of the room.

● Draw a picture of the room to show where the bottle is and where everyone is sitting.

Your teacher will take the top off the bottle.

● Mark on your drawing the order in which people notice the smell of the chemical.

1. Use the idea about the smell being made up of particles to explain the results of the experiment.

In this activity you will use the particle idea to make a prediction about air.

What you need

metre ruler
triangle-shaped support for ruler
2 balloons or small plastic bags
freezer-bag ties or small rubber bands

What you do

● Tie the 2 balloons to opposite ends of the ruler.
● Balance the ruler across the support so that it is level.
● Write down the measurement at which the ruler balanced.
● Take off one of the balloons and inflate it (**D**).
● PREDICT what will happen when you fix the blown-up balloon back on to the ruler. Discuss how the particle idea lets you make that prediction.
● TEST your prediction.

D

Rubber band
Metre ruler
Rubber band
Balloon

Observations

Record what happens to the ruler when the inflated balloon is put back on to the end.

Now try this . . .

Set up two measuring cylinders as shown in **E**.

E

What will the girls see when they mix the rice and peas?

Predict what volume measurement you will see when you mix together the dried pea "particles" and the rice "particles".

Test your prediction and record your observations.

Now repeat the mixing experiment, but this time use two liquids, 50cm³ of water and 50cm³ of either ethanol or propanone. What are your observations?

How does the pea-and-rice experiment help to explain your observations of the liquids when you mixed them? What do you infer about the shapes of the particles?

2. What can you infer about the balloons when the ruler is balanced?

3. What did you add to the balloon when you blew it up?

4. How does the idea that air is made of particles help to explain your observation?

5. Think of an inference that does NOT use the idea of particles to explain your observation. Describe it to your partner.

6. If the particle idea is the best one, draw what you think the particles look like inside the balloon:
   (a) before you blew into it;
   (b) after you inflated it.

7. How does the particle idea help to explain what happens if you let go of a balloon that has been inflated but not tied off?

ALL CHANGE

Read about the adventures of Mollie Cule and answer the questions that follow.

> Hi there! My name's Mollie Cule! There are millions like me in a drop of water! Millions of water Mollie Cules!

> All water molecules look the same. We are all made of three little atoms – two hydrogen atoms and one atom of oxygen. We're quite simple molecules really.

> Most people know me by my code name, H-two-O, usually written H_2O.

> Here I am in the team photograph! It's a bit of a cheat really, because I'm far too small to be seen by any sort of microscope. Scientists can only imagine what I look like because of what I do.

> Let me give you an idea of how small I am. If I stood in line with 100,000,000 that's one hundred million of my water molecule friends, we would just measure 1 centimetre.

> If you tried it with 100,000,000 of your friends, you would stretch for over 38,000 kilometres! That's nearly all the way around the Earth's equator!

At the moment, I'm feeling quite warm and lively. There's plenty of room to move around in. Yesterday, it was really cold. The temperature fell to below 0 degrees Celsius and I seemed to have very little energy left to move around. Rows and rows of water molecules like me just stood around in the freezer, shivering quietly. We were well and truly **frozen** ice.

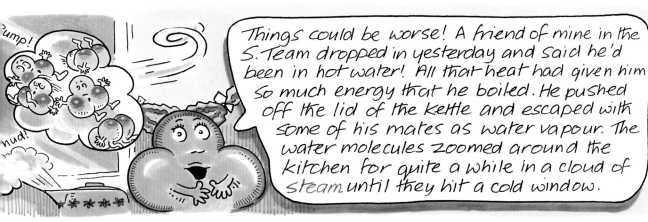

Things could be worse! A friend of mine in the S. Team dropped in yesterday and said he'd been in hot water! All that heat had given him so much energy that he boiled. He pushed off the lid of the kettle and escaped with some of his mates as water vapour. The water molecules zoomed around the kitchen for quite a while in a cloud of steam until they hit a cold window.

That really took the energy out of them, I can tell you. The S. Team condensed and slid down the window as a drop of liquid water, just waiting to be wiped up!

1. Write down one piece of information from the story that you think gives the best idea of the size of a water molecule.

2. Scientists say that there are three STATES of water. They are liquid water, solid ice water, and water vapour or steam.
Draw three pictures to show what the water molecules might look like in a small amount of each state.

3. Next to your pictures, write down two facts about the molecules in each state of water.

4. Suggest which states the water molecules might be in at these temperatures:
(a) 20°C
(b) -10°C
(c) 90°C
(d) 120°C

MELTING AND BOILING

When you heat a lot of substances they soften and melt. On further heating they start to evaporate and finally boil. Ice, for example, is solid (**A**), but if you warm it above 0°Celsius, it begins to melt and become liquid water. If you heat it more, the liquid water begins to evaporate until it boils at 100°Celsius.

These two temperatures, the **melting** and **boiling points**, are correct only for water. Other substances will have different values for melting and boiling.

The melting and boiling points of ten substances are given in Table 1. You will notice that substances that we usually think of as solids, liquids and gases can all exist in forms that melt or boil.

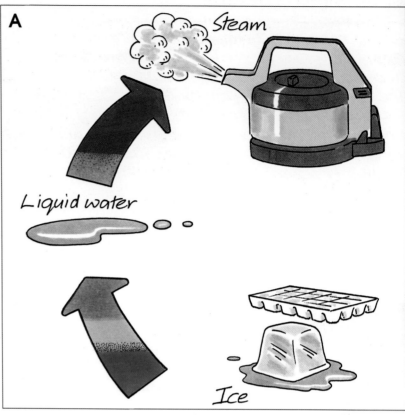

How does water change from one physical state to another?

Table 1

SUBSTANCES	BOILING POINT (°C)	MELTING POINT (°C)
Metals		
iron	2887	1539
gold	2707	1063
lead	1751	327
mercury	356	-39
Liquids		
water	100	0
ethanol	78	-117
propanone	56	-95
Gases		
methane	-162	-182
oxygen	-183	-219
nitrogen	-196	-247

Melting and boiling points can be used to indicate how pure a substance is. Only pure water melts at 0°C and boils at 100°C. Any other chemicals in the water will change those values.

What do you think will happen to the melting point of water (or freezing point - the temperature at which a substance goes from a liquid to a solid) if salt is added? Will it go up or down? What do you think will happen to the boiling point?

1. Draw bar graphs of the melting and boiling points against the names of the substances. Will you need to draw two separate graphs or will both sets of bars fit on one scale?

2. Which substance seems to be out of place on the graph? Give a reason for your choice. Describe what that substance looks like.

3. Why do you think mercury is used inside thermometers?

4. Give one advantage of using coloured ethanol inside a thermometer instead of mercury.

In these experiments, you will investigate the effect of salt on the melting point of ice and the boiling point of water.

What you need

2 small beakers
water
supply of crushed ice
2 thermometers
salt (sodium chloride)
weighing balance
clock or stop watch
Bunsen burner
tripod, gauze and heat-proof mat
stirring rod

 Wear goggles when heating. Do not stir liquids with a thermometer.

EXPERIMENT 1

What you do

- Set up the two small beakers as shown in **B**, one with ice only, the other with ice and salt.
- Think carefully about the variables you must control from the start.
- Measure the starting temperatures of the ice in each beaker.
- Start the clock and measure the time it takes for all the ice in each beaker to melt.
- Measure the temperature of the ice in each beaker at regular time intervals until the ice melts.

EXPERIMENT 2

What you do

- Using the apparatus available to you, design an experiment to find out if salt affects the temperature at which water boils.
- Make sure this test is as fair as Experiment 1.
- Ask your teacher to check your plan before you start the experiment.

5. What is the effect of salt on ice?

6. What inference can you make about the effect of salt on the freezing point of water?

7. Explain why salt is spread on roads in winter.

8. What inference can you make about the effect of salt on the boiling point of water?

Observations

In each experiment, you will have measurements of temperature and time. Decide on the best ways to display the numbers so that other people can see what has happened.

Now try this . . .

Water boils at different temperatures in different places. **C** shows the boiling points of pure water in four places.

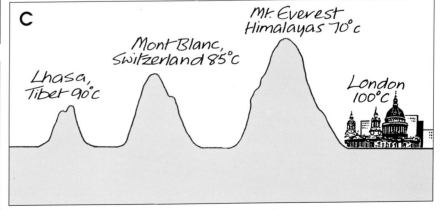

C — Mt. Everest Himalayas 70°C — Mont Blanc, Switzerland 85°C — Lhasa, Tibet 90°C — London 100°C

Write a sentence that links the boiling point of water to the height of a place above sea level.

ATOMIC BITS AND PIECES

Since the time of the "uncuttable atom" of Democritus, our ideas about the most basic parts of matter have changed and become more detailed. Those early thinkers did not test their ideas by experiment. It has only been in the last one hundred years or so that scientists have had the equipment to investigate substances on such a small scale and look for evidence to support their thought models for particles. Some of the models that modern scientists have of atoms are very complicated. However, the basic models (**A**) all seem to have some ideas in common. The atoms of nearly all substances seem to contain three sorts of smaller particle, **PROTONS**, **NEUTRONS** and **ELECTRONS**. You met these briefly in Book Two when you did some experiments on electricity.

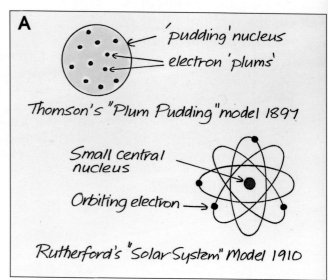

Which of these atom models seems most likely?

J.J. Thomson and Ernest Rutherford discuss their atom models.

PROTONS

The evidence suggests that these particles are in the centre of the atom, in a region called the **NUCLEUS**. The number of protons tells you what the substance is that the atom is part of. If an atom has one proton, the substance is hydrogen. If it has two protons, it is helium, if it has six protons, it is carbon, eight protons makes it oxygen, and so on. Clearly, the number of protons is very important and it is called the **ATOMIC NUMBER**. Protons have a positive electrical charge.

ELECTRONS

Electrons have a negative charge. Electrons fly around the nucleus much as planets spin around the sun. Electrons are much smaller than protons and neutrons. About 2000 of them would be needed to weigh the same as a proton. However, the electron's negative charge is equal to the positive charge on a proton and so in a neutral atom, with no overall charge, there are the same number of electrons as protons.

NEUTRONS

These particles share the nucleus with the protons. They do not have any charge, so they are neutral. Neutrons are about the same size and mass as protons, but their number in the nucleus can vary. Some carbon atoms have six neutrons, some have seven while others have eight. They are still carbon atoms because they have six protons.

If you add the number of protons to the number of neutrons, you get the **MASS NUMBER** of the atom. For example, oxygen with 8 protons and 8 neutrons will have a mass number of 8 + 8 = 16.

Table 1

Particle	proton	neutron	electron
Charge			
Where found			
Mass compared to an electron			

1. Copy Table 1 and fill it in with information from this page.

2. Draw a diagram of a helium atom which has 2 protons, 2 neutrons and 2 electrons.

C

Antoine Becquerel discovered radioactivity. Becquerel was a Frenchman and came from a family of physicists. He was very interested in Wilhelm Roentgen's recent experiments with X-rays and tried to find substances that would produce them.

D

In February 1896, Becquerel wrapped some photographic plates in black paper and put them in sunlight with a crystal of a chemical similar to one that Roentgen had used. Becquerel suspected that the crystal might produce X-rays under the effect of the sun. He predicted that the X-rays would go through the paper and affect the photographic plates. When he developed the plates, he found that they were fogged. Becquerel inferred that X-rays had produced the effect.

E

The weather turned cloudy and Becquerel could not carry on with his experiments. He had some fresh plates wrapped ready with crystals on top, but he could not use them. He put them in a cupboard and left them for about a month. For some reason, he decided to develop the plates and was surprised to see that they were fogged, just as before. Clearly, sunshine could not be involved. Becquerel now inferred that something different from X-rays had been sent out by the crystal. In 1898, following a similar set of experiments, Marie Curie named the effect "radioactivity".

F

By 1899, Becquerel had found that the radioactivity could be bent by a magnet. He inferred that some part of it must consist of tiny, charged particles. Then, in 1900, he discovered that the radioactivity behaved in the same way as the stream of speeding electrons made by passing electricity through a negatively charged wire. Later, he showed that it was the uranium in his crystals that produced the electrons. In 1903, Becquerel shared the Nobel prize for physics with Marie and Pierre Curie for his work on radioactivity.

---

3. In what way could Becquerel have improved his first experiment with the crystal and the photographic plates to make it a fair test?

4. Suggest a scientific reason why Becquerel decided to develop the plates that he had put in a cupboard.

5. Draw the apparatus that Becquerel might have used to show that a magnetic field could bend the stream of particles produced by a radioactive crystal.

6. An atom of uranium has 92 electrons and 143 neutrons. What is its atomic number? What is its mass number?

MarieCURIE 1867-1934

MARIE CURIE WAS BORN MARIE SKLODOWSKA IN WARSAW, WHICH WAS THEN UNDER RUSSIAN DOMINATION. HER MOTHER HAD DIED OF TB WHEN MARIE WAS QUITE YOUNG AND HER FATHER WAS A POOR SCHOOL TEACHER, SO IT WAS IMPOSSIBLE FOR HER TO HAVE A COLLEGE EDUCATION.

WHEN SHE LEFT SCHOOL. MARIE BECAME A GOVERNESS, EARNING ENOUGH MONEY TO FINANCE HER SISTER'S MEDICAL STUDIES IN PARIS. SHE TAUGHT HERSELF FROM ALL THE BOOKS SHE COULD GET HOLD OF — HOPING TO BE ABLE TO JOIN HER SISTER WHEN SHE HAD SAVED ENOUGH MONEY.

IN 1891, SHE WENT TO PARIS AND ENTERED THE SORBONNE. SHE WORKED FAR INTO THE NIGHT IN HER STUDENT'S QUARTER GARRET, LIVING OFF BREAD AND BUTTER AND TEA. ONCE SHE FAINTED FROM HUNGER IN A LECTURE.

SHE CAME FIRST IN THE PHYSICAL SCIENCE EXAMINATIONS IN 1893, AND BEGAN WORKING IN THE LABORATORY OF ONE OF HER LECTURERS, GABRIEL LIPPMAN. IN 1894, SHE MET PIERRE CURIE, THE FOLLOWING YEAR THEY WERE MARRIED. TOGETHER THEY WERE TO MAKE SOME VERY IMPORTANT DISCOVERIES.

IN 1896, BECQUEREL DISCOVERED THE EMISSION OF RAYS FROM URANIUM SALTS. THE DISCOVERY FASCINATED MARIE CURIE — SHE NAMED THE PHENOMENON RADIOACTIVITY AND DECIDED TO DO THE WORK FOR HER DOCTOR'S THESIS IN THIS FIELD.

MARIE CURIE DEVISED A METHOD FOR MEASURING THE INTENSITY OF RADIATION FROM A GIVEN MATERIAL, BASED ON THE FACT THAT RADIATION IONISED THE AIR, MAKING IT CAPABLE OF CONDUCTING A DETECTABLE ELECTRIC CURRENT. SHE FOUND THAT THE INTENSITY WAS PROPORTIONAL TO THE AMOUNT OF URANIUM PRESENT, BUT THAT SOME URANIUM MINERALS SEEMED TO BE FAR MORE RADIOACTIVE THAN COULD BE ACCOUNTED FOR BY ANY URANIUM CONTENT.

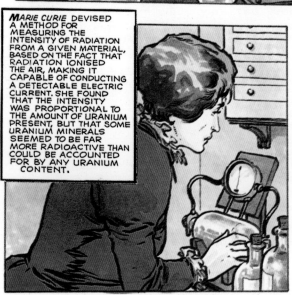

SHE DECIDED THAT THE ORES MUST CONTAIN ELEMENTS MUCH MORE RADIOACTIVE THAN URANIUM. HER HUSBAND GAVE UP HIS OWN RESEARCH AND TOGETHER THEY SET ABOUT FINDING THIS SUBSTANCE. IN 1898, THEY ISOLATED A NEW ELEMENT HUNDREDS OF TIMES MORE RADIOACTIVE THAN URANIUM. THEY CALLED IT POLONIUM, AFTER MARIE 'S NATIVE POLAND. EVEN POLONIUM DID NOT SEEM RADIOACTIVE ENOUGH AND THEY CONTINUED THEIR WORK. AT THE END OF 1898, THEY DETECTED A TRACE OF ANOTHER HIGHLY RADIOACTIVE SUBSTANCE, WHICH THEY CALLED RADIUM.

WHAT THE *CURIES* NEEDED WAS TO OBTAIN RADIUM IN LARGE ENOUGH AMOUNTS TO PROVE ITS PROPERTIES. THE SILVER MINES AT *JOACHIMSTHAL* IN *BOHEMIA* PRODUCED HUGE QUANTITIES OF URANIUM SLAG AND THE OWNERS ALLOWED THE *CURIES* TO TAKE AS MUCH AS THEY WANTED. THEY SPENT ALL THEIR SAVINGS IN SHIPPING IT TO *PARIS*.

THEY OBTAINED PERMISSION TO USE AN OLD, LEAKY, WOODEN SHED AT THE PHYSICS SCHOOL WHERE THEY BOTH WORKED. THERE, FOR FOUR YEARS, THEY REFINED THE ORE. *MARIE* HAD A SMALL DAUGHTER, BORN IN 1897, TO LOOK AFTER, BUT STILL SHE KEPT AT WORK. IN 1902, THEY HAD REFINED A TENTH OF A GRAM OF RADIUM.

IN 1903, *MARIE* WROTE HER DOCTOR'S THESIS AND, THAT YEAR, THE *CURIES* SHARED THE *NOBEL* PRIZE FOR PHYSICS WITH *BECQUEREL. PIERRE* BECAME A PROFESSOR AT THE SORBONNE THE NEXT YEAR BUT HIS CAREER THERE WAS TO BE SHORT. IN 1906 HE WAS RUN OVER AND KILLED BY A HORSE-DRAWN CARRIAGE.

MARIE CURIE TOOK OVER *PIERRE'S* POSITION AT THE SORBONNE, BUT PREJUDICE AGAINST WOMEN PREVENTED HER BECOMING A MEMBER OF THE FRENCH ACADEMY, ALTHOUGH SHE WAS NOMINATED. IN 1911, SHE RECEIVED THE *NOBEL PRIZE* IN CHEMISTRY FOR HER DISCOVERY OF THE TWO NEW ELEMENTS, POLONIUM AND RADIUM. SHE IS THE ONLY PERSON TO HAVE RECEIVED TWO *NOBEL* SCIENCE PRIZES.

THROUGHOUT THE FIRST WORLD WAR, *MARIE,* WITH THE HELP OF HER DAUGHTER *IRENE,* DEVOTED HERSELF TO THE DEVELOPMENT OF X-RADIOGRAPHY. LIKE MANY WOMEN, SHE HELPED THE WAR EFFORT BY DRIVING AN AMBULANCE.

IN 1921, *MARIE* WENT TO AMERICA, ACCOMPANIED BY BOTH HER DAUGHTERS. THERE, PRESIDENT *WARREN HARDING* PRESENTED HER WITH A GRAM OF RADIUM BOUGHT WITH MONEY COLLECTED BY AMERICAN WOMEN. SHE GAVE LECTURES ALL OVER THE WORLD DURING THE NEXT DECADE. LATER, *MARIE CURIE* SUPERVISED THE PARIS INSTITUTE FOR RADIUM. DESPITE THE CANCER-CURING PROPERTIES OF RADIUM, SHE HERSELF DIED OF LEUKEMIA CAUSED BY OVER-EXPOSURE TO RADIATION.

RADIOACTIVITY - GOOD OR BAD?

Some atoms lose bits of themselves when they are unstable. Usually, the larger and heavier the atoms are, the more unstable they are. This type of atomic "falling apart" is called **RADIOACTIVE DECAY**. There are three types of radiation that are commonly produced by radioactive materials (**A**).

1. **Alpha particles** - these each consist of 2 protons and 2 neutrons and can only travel a few centimetres from their source.

2. **Beta particles** - these are fast-moving electrons and can travel about 20cm in air or just a few centimetres through a hand.

3. **Gamma rays** - are a much more powerful type of radiation. They can pass through sheets of aluminium and lead and can only be stopped by a block of a material such as concrete.

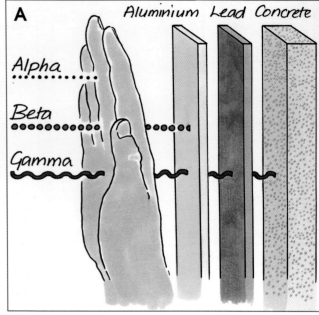

Types of radiation.

Some atoms fall apart naturally and are doing it all around us. This is one cause of what is known as **"background radiation"**. Some rocks, such as granite, contain elements that are constantly giving out a low level of radioactive particles.

Other atoms can be made to fall apart and the radiation as well as the energy they give out when they decay can be made to do work for us.

The unit used to measure doses of radiation is the **SIEVERT** (Sv). A dose of 10 Sieverts is used on a small part of the body to kill cancer cells. If the whole body was exposed to this level of radiation, it would be fatal. To measure less dangerous amounts of radiation, the **millisievert** (mSv) is used. One mSv is one thousandth of a Sievert.

B shows the amounts of radiation people in the UK receive naturally each year from various sources.

Sources of radiation in the UK. Which of these can be avoided?

Electricity can be generated from nuclear fuel.

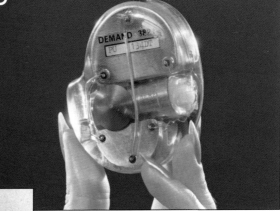

A heart can be kept beating with a nuclear battery.

Huge amounts of energy are released in a nuclear explosion.

The amount of material flowing along a pipe can be measured using radioactivity.

1. Look at the pictures and make sure you understand what job the radioactivity has to do. Pictures **C** - **F** show some uses of radioactivity.

2. For each use shown, decide whether you think it is a "good" or a "bad" use of radioactivity. Write down your reasons.

3. Discuss your ideas with your partner and then in a bigger group. Perhaps you could imagine that you are in a position to give money for research into uses of radioactivity. What uses would you give money to? What sorts of research would you want to stop? Why?

CHEMICAL CHANGES

When chemicals change into something new, scientists say that they **react**. Chemical reactions are everywhere. In the kitchen, chemical reactions help to make tasty meals. Bread is made by encouraging yeast microbes to grow and change **sugar** into **carbon dioxide** gas bubbles that make the bread dough rise. The conditions have to be just right for the changes to happen (**A**). Too cool and the reaction will not start. Too hot and the yeast will die.

Chemical reactions are at work in many industries. Pictures **A** - **F** show a few of them. There are jobs for scientists in all of these industries.

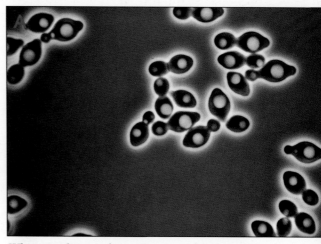

What conditions does yeast need to grow?

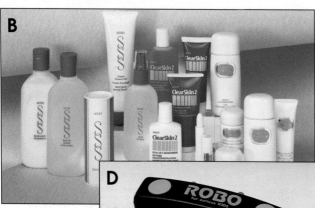

Cosmetics.

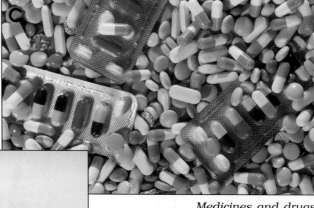

Medicines and drugs.

Plastics.

Fibres, fabrics and dyes.

Fuels.

1. Look around you now or think about the rooms at home.
   Make a list of 6 substances that may have been made by a reaction between different chemicals or that might take part in a reaction themselves. For each example, say what chemicals you think were used to make it or what it can be made into itself.

In this activity, you will be able to observe a variety of reactions and think about what happens in a reaction.

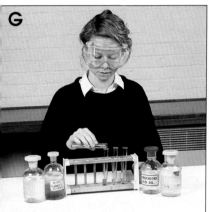

G

What you need

Solutions of these chemicals:
A sodium carbonate
B copper sulphate
C dilute hydrochloric acid
D dilute ammonium hydroxide

supply of test tubes
test tube rack

 Wear goggles when mixing solutions. Do not smell any solutions unless your teacher shows you the safe way to do so. Wash your hands after handling chemicals.

What you do

● Measure out about 1cm³ of each solution into a test tube.
● Add solution **A** to solution **B**, as shown in **G**.
● Look carefully at the mixed solutions to see if there is any reaction change.
● Can you use your other senses to make observations as well?
● Add solution **C** to solution **D** and repeat your observations.
● Wash out the test tubes and add 1cm³ of each solution again.
● Carry on mixing pairs of solutions until you have tried all the possible combinations.

Observations

H

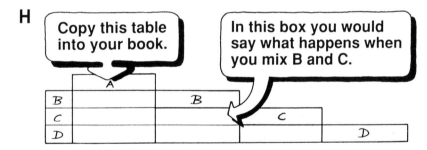

Copy this table into your book.

In this box you would say what happens when you mix B and C.

	A	B	C	D
B				
C				
D				

2. Which pairs of solutions seemed to react together? List them.

3. Write down one observation that you think best shows that a reaction change has happened.

4. Part of a page from Janita's note book is shown in **I**.
Read the inference she made about what was happening in one of the tubes.
Discuss her inference with your group. Could it be true? Are there any other possible inferences?

5. Try to write down other sentences, like Janita's "sums", to show what new things have been made in all the tubes where you observed reactions.

I

I think that when I poured the sodium carbonate solution into the copper sulphate solution, the chemicals mixed up to make new ones.

This sentence says what I mean. It's a bit like a sum:

Sodium Carbonate + Copper Sulphate

⇩

Sodium Sulphate + Copper Carbonate

Now try this . . .

Here is a list of changes. Some of them are **chemical** changes, where a new substance is made, while others are **physical** changes, not involving the making of new chemicals.
Say which changes are chemical and which are physical. Give reasons for your choices.

(a) Burning a piece of wood
(b) Sawing a piece of wood in half
(c) Running a car using a petrol engine
(d) Dissolving sugar in tea or coffee
(e) Baking a cake

ACIDS

Most people have heard the word "**acid**" in the last few years in discussions about "**acid rain**" and its effects on wildlife and the environment. In fact, acids are a common family of chemicals that share several properties. Scientists put them together because of the reactions that they can take part in.

All the chemicals shown in **A** are acids.

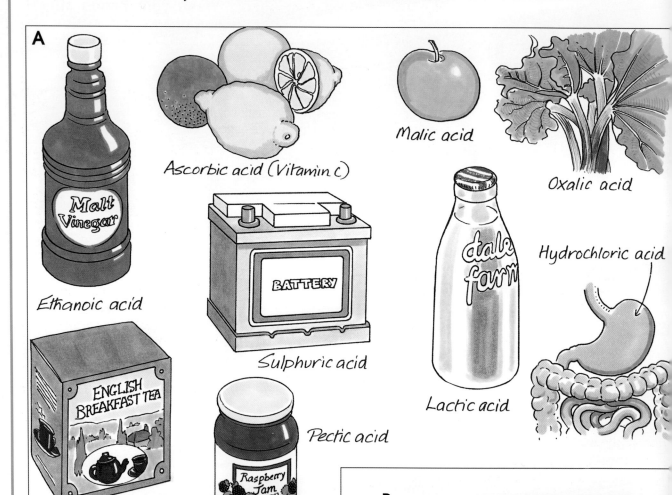

Acids in our lives.   Which of them are safety hazards?

Wherever acids are in use, you will see the sign in **B**. "**Corrosive**" means that it will produce chemical burns on lab benches, clothes or skin. Your teacher will not let you use very strong acids that have a high risk, but even diluted acids should be treated with great care.

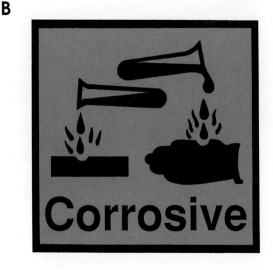

This experiment looks at the reactions between some acids and three common metals used in industry.

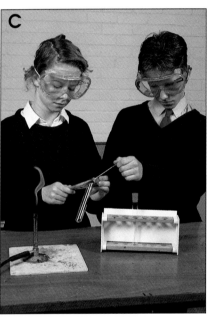

C

Testing gases.

1. Name the gas or gases that were made in the experiment.

2. List the metals in the order in which they make gas quickest.

3. List the acids in the order in which they make gas quickest.

4. Infer where the gas comes from. Can you test your inference by a further experiment?

5. Write at least one sentence or **word equation** that says what happened in one of the reactions.

What you need

small pieces of iron, copper and magnesium
small bottles of dilute:
ethanoic acid (vinegar)
sulphuric acid (in car batteries)
hydrochloric acid (found in our stomachs)
nitric acid (used to make fertilisers)

supply of test tubes
boiling tubes
wooden splint
lime water
gas collection equipment (see Pupil Book One "Gases in the Air")
"Gas Testing Sheet" (from Resource Book Three)

> ⚠ **Wear goggles and protective clothing when handling acids or using a Bunsen burner. Do not do this experiment if you suffer from breathing problems. Do not wash pieces of metal down the sink.**

What you do

● Place small amounts of the metals in your test tubes.
● Add a small amount of one of the acids to the metals, starting with ethanoic acid (**C**).
● Observe any reaction in the tube.
● If any of the mixtures seem to bubble and produce gas, try to collect it for 5 minutes by any method that your teacher tells you is **SAFE**.
● Follow the instructions on the "Gas Testing Sheet" to identify the gas produced in the reaction.
● If you have not tested gases before, your teacher will demonstrate the methods first.
● Repeat the experiment with all the metals in all the acids.

Observations

Make up a table to show what happened when you added each metal to each acid.

D

Cotton wool with dilute hydrochloric acid added.

Sandwich box and lid.

Pot of grass or cress seedlings.

Dish of sodium bisulphite (hydrochloric acid with sodium bisulphite makes sulphur dioxide gas.)

Sandwich box "greenhouse".

Now try this . . .

Sulphur dioxide is one of the "**acid gases**". This means that when it dissolves in water, it becomes an acid. It is thought to be one of the gases responsible for acid rain. Set up the "acid rain" experiment shown in **D**.

Investigate the effect of this acid rain on a dish of growing plants, such as grass or cress.
What changes can you observe?

ALKALIS AND NEUTRALISATION

You have seen that there is a great variety of acids, many of them found naturally in fruits as well as the mineral acids that you may be more familiar with, such as **sulphuric**, **hydrochloric** and **nitric** acids. Most of them are able to produce **hydrogen gas** when they react with a **metal**, although some of them react very slowly.

There are many situations where it is important to be able to remove or **neutralise** the effect of an acid, which might otherwise damage its surroundings. For example, if there is too much hydrochloric acid in your stomach, it could irritate the stomach lining, "eating" it away to make an **ulcer** (**A**). Chemists are working hard to produce medicines that will fight the effects of the acid.

Many people use a chemical called **sodium bicarbonate** as a stomach powder. It reacts with the acid to make three harmless chemicals. These are **water**, **carbon dioxide gas** and common salt (**sodium chloride**).

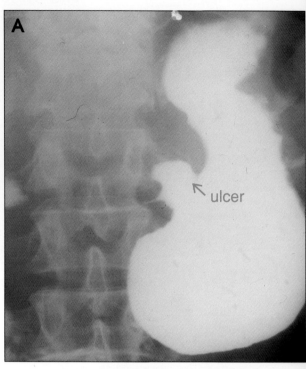

How has this special sort of photograph been taken?

---

hydrochloric acid + sodium bicarbonate ⇨ water + carbon dioxide + sodium chloride (salt)

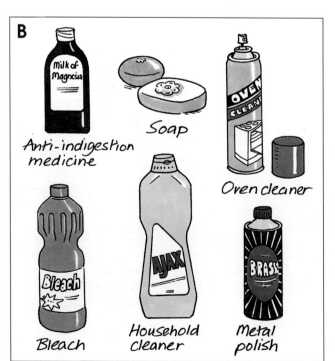

Uses of alkalis.

Sodium bicarbonate belongs to a family of related chemicals called **alkalis**. Alkalis are able to neutralise acids and get rid of their corrosive effects. By themselves, strong alkalis can cause burns similar to the ones produced by strong acids, so you have to take care when using them. Alkalis are used widely in the manufacture of important chemicals, generally called **salts**, by neutralisation.

A sentence that describes most neutralisation reactions is:

acid + alkali ⇨ water + a salt

Some uses of alkalis are shown in **B**.

In this investigation, you will use the technique called titration to find out the strength of an acid by neutralising it with an alkali of known strength.

The chemists who are testing new stomach medicines need to measure the strength of stomach acid accurately when they neutralise it with an alkali. To help them to do this, they add a coloured chemical called an **indicator** to the mixture. Indicators change colour when the strength of the acid changes.

What you need

10cm³ syringe
25cm³ burette or large syringe
small beaker or conical flask
supply of alkali with strength 2 Molar (2M)
supply of acid of unknown strength
bottle of liquid indicator
filter funnel

! Wear goggles when using acids or alkalis.

What you do

- Use the small syringe to measure 10cm³ of the alkali.
- Let the alkali run into a small beaker or conical flask.
- Add a few drops of the indicator liquid to the alkali (**C**).
- Fill the burette or a large syringe with the acid (**D**).
- Write down the volume of acid in the burette or syringe.
- Add the acid very slowly to the alkali, almost a drop at a time.
- Gently swirl the beaker to mix the acid and alkali.
- Keep adding acid until the mixture is a pale green colour.
- Write down the volume of acid left in the burette or syringe.
- Repeat the experiment at least once more.

C

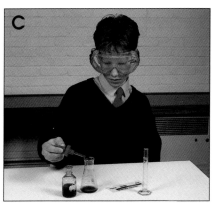

D

HOW STRONG IS YOUR ACID?
If you used more than 10cm³ of acid to neutralise 10cm³ of alkali, the acid must be weaker than the alkali.

For example:
If 20cm³ of acid neutralised 10cm³ of 2M alkali, the strength of the acid is:
$$\frac{10}{20} \times 2M = 1M$$

Observations

E

Copy this table into your book.

Burette readings (cm³)			
	At start	At end	Volume of acid used
1st go			
2nd go			
		Average:	

1. Write a sentence that explains what each of these words means:
   (a) neutralisation
   (b) indicator
   (c) titration

2. Using the average volume you have measured, work out the strength of the acid you were given.

3. The acid in your stomach is about 2M. The volume of acid in your stomach is 20cm³ when you are digesting some food. Which of the following amounts of alkali medicine would just neutralise the acid:
   (a) 10cm³ of 1M alkali?
   (b) 20cm³ of 2M alkali?
   (c) 40cm³ of 1M alkali?

INDICATORS

Substances that change colour when you put them into acids or alkalis are called **indicators**. Indicators are often made from natural, living things, such as flowers, fruits or vegetables, although some are now chemically produced in the laboratory. Even the balance of acid and alkali in the soil in which some plants grow can affect the colour of their flowers, such as hydrangeas (**A**, **B**).

Hydrangea in acidic soil.

Hydrangea in alkaline soil.

In the last Unit, you may have used a chemical called Universal Indicator to show when a mixture of an acid and an alkali has been neutralised. Universal Indicator can have several different colours, depending on how acidic or alkaline the solution is that it has been added to.

You can compare the colours you see with colours on a special chart. This is called a **pH** ("pee-aych") **chart**. Scientists made up the pH scale as a way of saying how strongly acidic or alkaline a substance is. **C** shows what the numbers on the pH scale mean together with the pH of some common substances. **D** is a photograph of the different colours that Universal Indicator can change to.

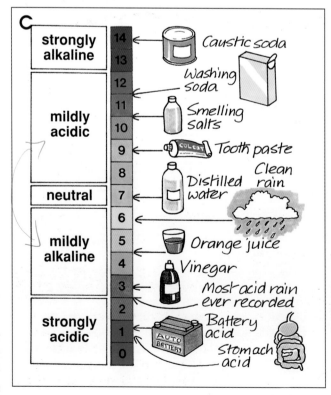

The pH scale indicates whether a substance is acidic, alkaline or neutral.

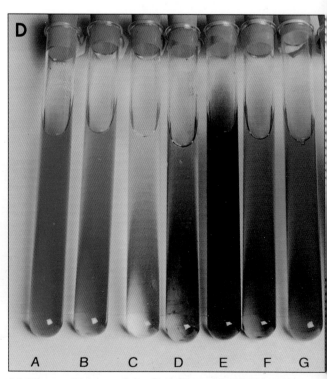

*Match a substance from **C** with each of these tubes.*

In this activity, you can make your own indicators from natural things. You will be able to investigate how well they indicate the pH of acids and alkalis.

What you need

natural materials to make indicators
mortar and pestle
Universal Indicator
samples to pH test
dropping pipettes
supply of distilled (pure) water
supply of test tubes and rack
clean watch glasses or plastic lids

Wear goggles if handling chemicals.
Wash your hands after handling chemicals.
Do not wash solid substances down sinks.

What you do

- Use the apparatus provided to produce about 10cm³ of juice extract from one of the natural materials (**E**). This is your own indicator.
- Add a few drops of your indicator to small samples of various common substances on watch glasses or lids (**F**).
- If the substance being tested is a solid, you might need to add a few drops of water to it first.
- Wash the lids clean with distilled water.
- Repeat the tests with Universal Indicator.

E

Observations

G

Substance	Colour with my indicator	Colours with Universal Indicator	pH	Acid, alkali or neutral?
Milk				

F

Indicator liquid · milk · Milk · Universal Indicator

1. Why is it important to wash the lids clean with distilled (pure) water before you do another pH test?

2. If you made more than one indicator for yourself, name the one that is best at indicating different strengths of acid and alkali. Explain the reason for your choice.

3. How does your best indicator compare with Universal Indicator?

4. Would any of the juice extracts you have made be suitable as an "invisible" ink? What would you have to do to make the writing visible?

Now try this . . .

Design an experiment that could find out whether the dyes used in sweets, such as Smarties, are sensitive to pH. If possible, carry out your test and report back to the class.

NEW COMBINATIONS

When you burn a piece of coal (**A**), atoms of the element carbon in the coal join up with molecules of oxygen, in the air, to make molecules of the gas carbon dioxide. The word equation for this reaction would be:

$$\text{carbon} + \text{oxygen} \Rightarrow \text{carbon dioxide}$$

You can't easily break down the carbon dioxide to make carbon and oxygen again. This is an example of a chemical change and scientists call this sort of reaction a **combination** or **synthesis** reaction. Whenever fuels are burned, millions of combination reactions happen between carbon and oxygen as carbon dioxide is synthesised.

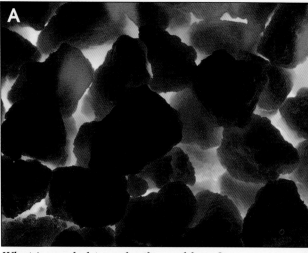

What is needed to make the coal burn?

Pictures **B** and **C** show two more chemicals that are going to take part in a combination reaction. These are the **elements** iron and sulphur. Iron is a metal and sulphur is a non-metal. **D** shows what is made when the iron is mixed with the sulphur and the two elements are heated together. The combination reaction between them has synthesised a new chemical or **compound** called iron sulphide:

$$\text{iron} + \text{sulphur} \Rightarrow \text{iron sulphide}$$

Iron filings.

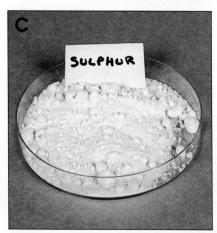

Sulphur powder.

Iron sulphide.

SUMMARY

1. When pure **elements** are mixed together, such as iron and sulphur, scientists call that a **mixture**.

2. When elements are mixed together and made to react, the new chemical that is made is called a **compound**, such as iron sulphide.

3. The **elements** that have been combined to make a compound cannot be easily separated into their original form.

1. What observations can you make of the chemicals involved in the reaction to suggest that a chemical change has taken place?

2. When some students carried out this combination reaction with iron and sulphur, one of them asked for a magnet. The student said that she could use it to find out if something new had been made. Explain why a magnet could be used to find that out.

In this activity you will be able to find out more about combination reactions and compounds.

What you need

small piece of clean copper metal foil
Bunsen burner and heat-proof mat
tongs
accurate weighing balance

 Wear goggles when heating. Never weigh hot objects.

What you do

● Weigh your piece of copper foil as accurately as you can. Ask for help if you cannot read the balance.
● Hold the foil in the tongs and heat it strongly for 5 minutes (**E**).
● Let the foil cool and weigh it again.

Observations

Record your observations and measurements to describe what happened to the copper when you heated it.

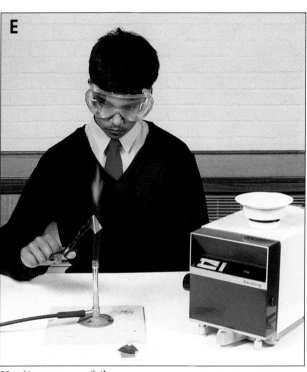

E

Heating copper foil.

F

The carbon block weighed five grams when I started to heat it. After one minute, it weighed four grams. Nine minutes later, it weighed one gram. By twenty minutes, all the carbon had disappeared!

Arifa's measurements.

3. Explain the meaning of these science words:
   (a) combination reaction
   (b) element
   (b) mixture
   (c) compound

4. What is the evidence that suggests that the copper has changed from a pure element into a new compound?

5. Compare your measurements with other groups in your class. What reason can you give to explain any differences?

6. Suggest what element might have combined with the copper to make the new compound. Can you give the compound a name?

Now try this . . .

Arifa heated a block of carbon for 20 minutes and weighed it at various intervals. She is describing her results in **F**.

Infer a reason for the loss of mass.

Name the new compound synthesised in Arifa's experiment.

Suggest how you might collect the new compound and identify it.

HOT BREAKDOWN

When you make toast you are performing a chemical reaction! **A** shows some bread that has been toasted. Bread contains compounds called **carbohydrates**, basically made from atoms of the element carbon joined to molecules of water. The effect of the heat is to break the link between the carbon and the water. The water is driven off as steam, leaving black carbon behind. That's toast!

The sort of reaction where heat energy is used to break down larger compounds into smaller, simpler ones is called **thermal decomposition**.

What chemical change has happened here?

An historic example of a thermal decomposition reaction is the one carried out by the scientist Joseph Priestley in 1774. Priestley was the first person to make oxygen gas by heating mercury in air and then breaking down the mercury compound that was made back to mercury metal again plus oxygen gas (**B**).

Priestley's research was important in helping the French scientist, Antoine Lavoisier, to identify oxygen as the gas that is needed for burning to take place.

Joseph Priestley.

ANTOINE LAURENT LAVOISIER,
FERMIER GÉNÉRAL NÉ A PARIS LE 16 AOUT 1743.

Antoine Lavoisier.

The word equations for Priestley's reactions look like this:

Reaction 1: A combination reaction

| mercury + oxygen in air $\Rightarrow$ mercury oxide |

... then with more heating under a magnifying glass ...

Reaction 2: A thermal decomposition reaction

| mercury oxide $\Rightarrow$ mercury + oxygen |

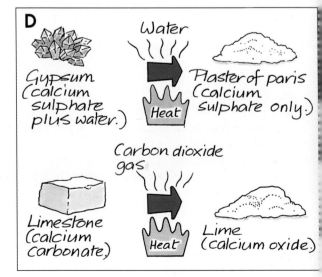

Some other thermal decomposition reactions.

In this experiment, you will investigate the gas given off when the compound copper carbonate is decomposed by heat.

What you need

sample of copper carbonate
supply of test tubes
dropping pipette
"Gas Testing" sheet from your teacher
lime water
wooden splints
Universal Indicator and pH colour chart
Bunsen burner and heat-proof mat
test tube holder and rack

Wear goggles when heating. Wash your hands after handling chemicals. Do not wash chemicals down the sink.

What you do

- Heat a small sample of copper carbonate slowly in a test tube.
- As you heat the compound, use the pipette to suck up some of the gas (**E**).
- Using the "Gas Testing" sheet as a guide, test the sample of gas to identify it.
- You might decide that you need to change the way you collect the gas. If you do, show your teacher a new plan before you change the method.
- Collect a sample of the chemical that is left in the test tube after heating.

results of the gas tests you performed.

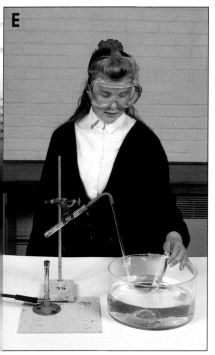

Collecting gas from the tube of copper carbonate.

Heating ammonium chloride crystals.

Now try this . . .

A compound that decomposes in an unusual way when it is heated is ammonium chloride.

Gently heat a small sample of ammonium chloride in a boiling tube and describe the changes that take place (**F**).

Infer a reason for the changes you observe.

Scientists say that ammonium chloride **sublimes**. Find out the meaning of the word "sublimes". Does your research support your inference in question 2?

1. What is the evidence that there has been a chemical change to the copper carbonate?

2. From your gas tests, name the gas produced when copper carbonate decomposes.

3. Copy and complete this word equation to describe the thermal decomposition reaction you have performed:
copper carbonate → .... gas + black ....

4. Draw a line graph to show how you would expect the mass of a test tube of copper carbonate to change during 10 minutes of heating. Label each axis clearly.

CHANGING PARTNERS

The scientist in **A** is mixing two different colourless solutions. **B** shows what she observes as soon as the solutions mix. She didn't need to heat them - they reacted straight away! Ten minutes later, the contents of the tube look like **C**. The white chemical sinks to the bottom of the test tube as a powdery layer or **precipitate**. It is insoluble in water.

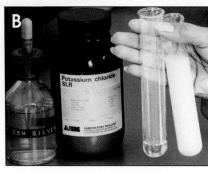

Scientists call this sort of reaction a **double decomposition** reaction, because both of the starting compounds have broken down into something new. Double decomposition reactions are important in the chemical industry when scientists want to make insoluble compounds.

Let's take a closer look at what happened in the test tubes. The starting chemicals were silver nitrate and potassium chloride. Scientists believe that you can think of the starting chemicals as ballroom dancers (**D**). When the chemicals swapped partners, silver chloride sank to the bottom of the tube because it does not dissolve in water.

| silver nitrate + potassium chloride ⇨ silver chloride + potassium nitrate |

1. Copy and complete these double decomposition reactions. Reaction (a) has been done for you.
   - (a) sodium chloride + lead nitrate → sodium nitrate + <u>lead chloride</u>
   - (b) copper sulphate + potassium carbonate →
   - (c) calcium chloride + sodium carbonate →
   - (d) sulphuric acid (hydrogen sulphate) + lead nitrate →
   - (e) magnesium sulphate + sodium carbonate →
   - (f) copper nitrate + potassium carbonate →
   - (g) zinc nitrate + sodium carbonate →
   - (j) copper sulphate + potassium hydroxide →

2. Use the Factbox **E** to predict which of the compounds (**salts**) made when the compounds react will be the insoluble one. Underline it in your book. The first one has been done for you.

In this activity you will be asked to predict the products of some double decomposition reactions and then test your predictions.

What you need

6 test tubes and rack
solutions of:
  copper sulphate
  sodium carbonate
  iron sulphate
  sodium hydroxide
  lead nitrate

What you do

● Copy table **F** and predict the result of mixing each pair of solutions.
● Fill the 6 tubes to a depth of 1cm. with the solutions.
● Add 1cm. of either sodium carbonate or sodium hydroxide solution to the tubes. Use Table **F** as a guide.

Wear goggles and protective clothing when mixing chemicals.
Wash your hands after handling chemicals.

E **SOLUBILITY FACTBOX**

● ALL salts of sodium and potassium are soluble.
● ALL nitrates are soluble.
● MOST chlorides are soluble, except silver, lead and barium.
● MOST carbonates are insoluble except sodium and potassium.
● MOST lead salts are insoluble except lead nitrate.
● MOST hydroxides are insoluble except sodium, potassium and ammonium.

Observations

F

Copy this table into your book.

	Copper Sulphate Solution	Sodium carbonate Solution	Iron Sulphate Solution
Sodium hydroxide Solution			
Lead nitrate Solution			

Write down what you see when you mix the solutions.

G

3. How many of the reactions support your predictions? Give yourself a mark out of six!

4. Write word equations for each of the six reactions you have performed.

5. Using the information in Factbox **E**, underline which of the compounds made in the reaction is insoluble.

6. Suggest how you could separate the insoluble compounds that have precipitated from the other solutions. Draw a diagram of the apparatus you would need.

Now try this . . .

G shows an industrial version of your test tubes! Chemical reactions go on in the tall tank on the right.

List 3 differences between the industrial version and your test tube experiment.

Say how each difference is important in making commercial amounts of compound.

REACTIVITY

Displacement is a further sort of reaction that scientists use to make new chemical compounds. **A** and **B** show this sort of reaction at work. A clean iron nail is placed in a solution of copper sulphate and left for 10 minutes. Can you see why it is called a "displacement" reaction? What is being displaced? Infer the reaction taking place in the beaker.

Before ...

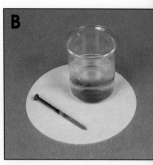
... and after.

You can also think of displacement as being like dancing! A couple who are dancing together are joined by a third person (**C**). If this "third person" is much stronger and more determined than one of the dancers, that dancer will be pushed out and replaced by the "third person"!

Scientists say that the chemical "dancer" that displaces another is more **reactive** than the "dancer" it displaces. **Reactivity** is the ability of one chemical element, usually a metal, to displace another from a compound.

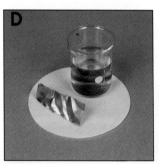
Before ...

... and after.

1. In the reaction between iron and copper sulphate solution, which is the more reactive metal, the iron or the copper? Give a reason for your choice.

2. Look at **D** and **E**. These pictures show what happens when a piece of copper is put into a solution of iron sulphate. Do these pictures support your answer to question 1? If they do, explain why. If they do not, what new inference can you make about the reactivity of the metals?

Scientists have performed many displacement reactions and can now list the metals in order of their reactivity. They call this list the **reactivity series** (**F**).

They have been able to place hydrogen, a non-metal element, on the list as well. It is useful to have hydrogen in the list because metals above hydrogen in the series will displace it from an acid. You can think of sulphuric acid, for example, as "hydrogen sulphate" and hydrochloric acid as "hydrogen chloride".

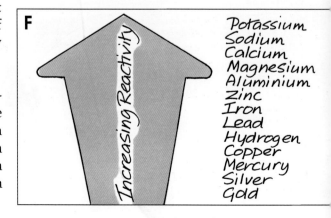

Increasing Reactivity

Potassium
Sodium
Calcium
Magnesium
Aluminium
Zinc
Iron
Lead
Hydrogen
Copper
Mercury
Silver
Gold

This investigation asks you to use the reactivity series to predict what will happen when metals are added to various solutions.

What you need

dilute solutions of the following:
   copper sulphate
   silver nitrate
   sulphuric acid
clean copper wire
iron or metal scrap
magnesium metal strip
supply of test tubes and rack

> ⚠ **Wear goggles when handling chemicals. Wear protective clothing. Wash your hands after handling chemicals. Do not try to wash metals down the sink.**

What you do

- Use the reactivity series (**F**) to predict what will happen (if anything) when you add the following metals to the solutions:
     copper + sulphuric acid
     copper + silver nitrate
     iron + copper sulphate
     magnesium + copper sulphate
     magnesium + sulphuric acid
     iron + sulphuric acid
     iron + silver nitrate
     magnesium + silver nitrate
- Use the apparatus given to you to plan an experiment to investigate the reactions.
- Ask your teacher to check your plan for safety.
- Carry out your plan using small amounts of all metals and solutions.
- You might like to bend the wire into different shapes first!
- Remember not to wash metals down the sink.

G

Iron with copper sulphate.

H

Copper with silver nitrate.

I

Magnesium with sulphuric acid.

Observations

J
> ⇨ zinc chloride + hydrogen
> ⇨ copper + zinc sulphate
> ⇨ silver + copper sulphate
> ⇨ iron sulphate + hydrogen gas

Su Lim's notebook.

3. Say how well you predicted the reactions. Were there any unexpected reactions? If so, suggest whether the reactivity series needs to be changed in any way.

4. Write down word equations to describe the displacement reactions you observed.

5. **J** shows some word equations from a set of experiments that Su Lim carried out. Unfortunately, the page from her notebook has been torn. For each reaction Su Lim observed, name the materials that she mixed together at the start.

> **Now try this . . .**
>
> How could scrap iron be used to produce pure copper from an impure solution of copper sulphate?
>
> What industry could make use of this reaction?

HENRY CAVENDISH 1731–1810

HENRY CAVENDISH WAS BORN IN *NICE*, WHEN HIS MOTHER, WIFE TO THE *THIRD DUKE OF DEVONSHIRE*, WAS ON HOLIDAY THERE TO IMPROVE HER HEALTH. HE WENT TO SCHOOL IN ENGLAND AND WENT ON TO *CAMBRIDGE* IN 1749. HE LEFT WITHOUT A DEGREE *(HE WAS AFRAID OF EXAMS)* BUT WITH A REPUTATION FOR BRILLIANCE AND ECCENTRICITY.

HE WAS KEPT ON A SMALL ALLOWANCE BY HIS FATHER AND HAD TO DO HIS EXPERIMENTING IN A SMALL LABORATORY THAT HE BUILT ABOVE THE STABLES OF THE FAMILY HOUSE IN GREAT MARLBOROUGH STREET, LONDON.

WORKING IN HIS LABORATORY *CAVENDISH* MADE DISCOVERIES, PARTICULARLY IN ELECTRICITY, THAT ANTICIPATED MUCH OF THE WORK OF *FARADAY* AND *COULOMB*. HOWEVER, HE PUBLISHED VERY LITTLE. HIS METHODS WERE DIRECT, AS HE HAD NO TALENT FOR INVENTING APPARATUS. HE MEASURED THE STRENGTH OF AN ELECTRIC CURRENT BY SHOCKING HIMSELF.

IN 1766, *CAVENDISH* MADE THE RESULTS OF SOME OF HIS RESEARCHES WITH 'AIRS' KNOWN TO THE *ROYAL SOCIETY*. HE HAD WEIGHED *'INFLAMMABLE' AIR* — WHAT WE NOW CALL *HYDROGEN* — AND FOUND IT TO BE VERY LIGHT. BECAUSE OF THIS, AND ITS INFLAMMABILITY, HE THOUGHT IT WAS PHLOGISTON.

WHEN HIS FATHER DIED IN 1783, *CAVENDISH* INHERITED OVER £1 MILLION. HE MOVED TO A HOUSE IN SOHO AND SET UP A LIBRARY THERE THAT THE PUBLIC COULD USE. BUT HE FOUND THE PEOPLE COMING IN A NUISANCE AND MOVED AGAIN, TO *CLAPHAM COMMON*, LEAVING HIS LIBRARY BEHIND. HE HAD TO FILL IN A FORM TO BORROW HIS OWN BOOKS.

HE BECAME EXCESSIVELY SHY AND ECCENTRIC. HE SAW HIS BROTHER AND HEIR, *FREDERICK*, ONLY ONCE A YEAR AND HIS ONLY SOCIAL CONTACT WAS THE REGULAR THURSDAY DINNER AT THE *ROYAL SOCIETY*. ON THE RARE OCCASIONS THAT HE ENTERTAINED, HE ALWAYS GAVE HIS GUESTS THE SAME THING, A LEG OF LAMB.

CAVENDISH DISLIKED WOMEN INTENSELY. HE COMMUNICATED TO WOMAN SERVANTS BY NOTES AND HAD A SPECIAL STAIRCASE AND SEPARATE DOOR BUILT FOR THEM. HOWEVER HE WAS COURTEOUS. HE ONCE SAVED A WOMAN FROM A COW THAT WAS CHASING HER ACROSS CLAPHAM COMMON.

CAVENDISH CARRIED ON WITH HIS EXPERIMENTS AT HIS CLAPHAM COMMON HOME. HE BUILT AN OBSERVATORY ON THE FIRST FLOOR AND A FORGE ON THE GROUND FLOOR. HE BEGAN EXPERIMENTING WITH AIR AND FOUND THAT A SMALL PART CONSISTED OF A VERY UNREACTIVE GAS THAT WE NOW CALL ARGON. HE DISCOVERED THE COMPOSITION OF NITRIC ACID BY FIRING A MIXTURE OF HYDROGEN AND AIR WITH AN ELECTRIC SPARK, AND GETTING WATER WITH NITRIC ACID DISSOLVED IN IT.

CAVENDISH PERFORMED HIS MOST SPECTACULAR EXPERIMENT IN 1798. HE CALCULATED THE MASS OF THE EARTH. THE PROBLEM WITH THIS HAD BEEN IN FINDING THE GRAVITATIONAL CONSTANT. HE DID THIS BY SUSPENDING A ROD WITH A LIGHT LEAD BALL AT EACH END BY A WIRE ATTACHED TO THE CENTRE. A LIGHT FORCE APPLIED TO THE BALLS MADE THE ROD TWIST. WHEN HE BROUGHT TWO LARGE BALLS NEAR THE SMALL ONES, THE FORCE OF GRAVITY BETWEEN THE LARGE AND SMALL BALLS TWISTED THE ROD. FROM THE AMOUNT OF TWIST, CAVENDISH CALCULATED THE GRAVITATIONAL CONSTANT. HE THEN WORKED OUT THE MASS OF THE EARTH, MAKING IT 6,600 MILLION MILLION MILLION TONS.

CAVENDISH DIED IN 1810. IT WAS A VERY SEVERE WINTER AND HE WENT OUT IN THE SAME CLOTHES THAT HE ALWAYS WORE. HE CAUGHT COLD WHICH DEVELOPED INTO PNEUMONIA. MUCH OF HIS WORK WAS NOT PUBLISHED UNTIL A CENTURY AFTER HIS DEATH, WHEN HIS NOTES WERE DISCOVERED BY MAXWELL.

MAKE HYDROGEN BY POURING DILUTE SULPHURIC OR HYDROCHLORIC ACID ON IRON FILINGS IN A BOTTLE (YOU CAN USE THE ACID FROM A CAR BATTERY). BE VERY CAREFUL AS HYDROGEN IS VERY INFLAMMABLE. FILL A BALLOON WITH THE GAS THAT IS GIVEN OFF. THE BALLOON WILL RISE STRAIGHT UP BECAUSE THE HYDROGEN IS SO MUCH LIGHTER THAN AIR.

OXIDATION AND REDUCTION

When a space rocket blasts off (**A**), fuel is being burned to release lots of energy. As you saw in **Book One**, when substances burn, oxygen is used up and water is made, together with the gas carbon dioxide. Whenever a fuel reacts with oxygen, scientists say that it is **oxidised**. Methane, for example, is a valuable fuel found in great quantities as natural gas (**B**). Scientists are closely involved in the search for natural gas.

The word equation for burning methane looks like this:

> **methane + oxygen** $\Rightarrow$ **water + carbon dioxide**

All natural fuels contain the element carbon combined with atoms of hydrogen. When methane burns, the carbon is oxidised to carbon dioxide and the hydrogen is oxidised to water.

Just as we can say that the methane is oxidised, we can say that the oxygen has been changed as well. It has been joined to hydrogen atoms to make water molecules.

This addition of hydrogen atoms to a chemical is called **reduction**. Scientists say that the oxygen has been reduced to water.

What chemicals are reacting here?

How do scientists and engineers help in the search for gas?

SUMMARY

1. **Oxidation** is the addition of oxygen to a chemical or the removal of hydrogen from it by an **oxidising agent**.

2. **Reduction** is the addition of hydrogen to a chemical or the removal of oxygen from it by a **reducing agent**.

3. **Oxidation** and **reduction** always happen at the same time - you can't have one without the other!

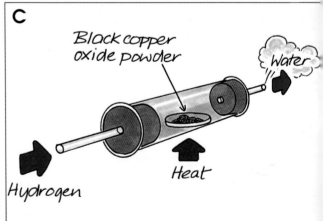

Heating copper oxide in hydrogen.

Look at this word equation for a different reaction (**C**). It is also an oxidation /reduction reaction, but it is not the same as when you burn methane.

copper oxide + hydrogen $\rightarrow$ **copper + water**

1. What is the big difference between this reaction and the one in which methane is burned?

2. Which chemical has been oxidised? Copper oxide or hydrogen?

3. Which chemical has been reduced? Copper oxide or hydrogen?

In this activity, you will investigate some unknown chemicals to see if any of them can take part in oxidation/reduction reactions.

A supermarket food manufacturer has sent you six solutions. One of them is labelled "**purple oxidising agent**" and one is labelled "**orange oxidising agent**". The other four tubes are labelled **A**, **B**, **C** and **D**. In the manufacturer's notes, it says:

"The purple solution loses its colour when mixed with a reducing agent. The orange solution changes to green when mixed with a reducing agent."

The food manufacturer wants you to find out whether any of the solutions **A**, **B**, **C** or **D** is a reducing agent.

 Wear goggles and protective clothing when handling chemicals. Wash your hands after handling chemicals.

What you need

sample of purple oxidising agent
sample of orange oxidising agent
samples of solutions **A**, **B**, **C** and **D**
dropping pipettes for each solution
test tubes and rack

What you do

- Using the apparatus given to you, plan your investigation to find out if any of the solutions **A**, **B**, **C** or **D** will reduce the coloured oxidising agents.
- Only use small amounts of the purple and orange solutions.
- Show your plan to your teacher before you collect any solutions.
- Carry out your tests.

Observations

Copy table **D** and record what happens in each tube. Do not ignore tubes in which nothing seems to happen!

D

	Solution A	Solution B	Solution C	Solution D
With purple agent				
With orange agent				

E

4. Write a sentence that explains what you understand by the words "oxidising agent" and "reducing agent".

5. From your observations, infer which of the four solutions **A** - **D** are reducing agents.

6. The food manufacturer uses the chemicals you have identified as "anti-oxidants" and adds them to crisps and biscuits. Suggest what job anti-oxidants might do.

Now try this . . .

Some household bleaches, such as "Domestos" are strong oxidising agents (**E**). From this information, suggest what an important job of oxidising agents might be.

ELECTRICITY AND REACTIONS

In the Unit "Hot Breakdown", you saw how heat energy can be used to break the chemical links that hold compounds together. Other sources of energy can also be used to make new substances. Electricity, for example, is a form of energy that can be used to decompose larger compounds into their basic particles. This process, in which electricity is used to help compounds to break down, is called **electrolysis**.

A shows the idea behind electrolysis. Many compounds, such as sodium chloride, common salt, are held together by an attraction between atoms that carry positive and negative charges. These charged particles are called **ions**. If you make a solution from one of these compounds and pass electricity through it, the ions can be made to separate. The positive ions move towards the negative side of the electricity supply, while the negative ions move towards the positive side.

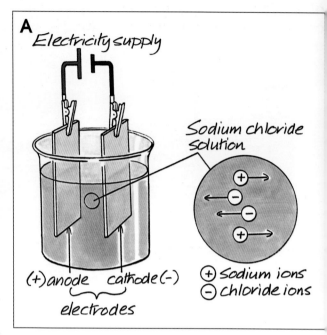

How does electricity help in this reaction?

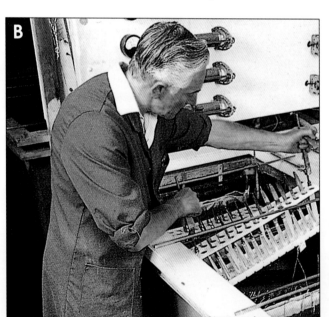

Silver-plating of cutlery.

In **B**, electrolysis is being used to coat steel cutlery with a layer of silver. The solution in the bath contains silver. Electricity is passed through it between racks of plain steel cutlery, acting as the negative **cathode**, and bars made from silver metal, which behave as the positive **anode**. The effect of the electrical energy is to force silver ions away from the bar of silver.

Since the silver metal ions are positively charged, they are attracted through the solution towards the steel cutlery cathode. When they reach the cutlery, the silver ions lose their charge and form a thin coat of pure silver metal (**C**).

This technique is known as **electro-plating**. Cutlery made in this way is marked **EPNS**, meaning "Electro-Plated Nickel Silver".

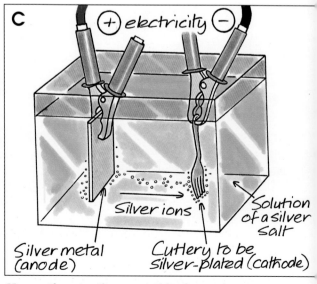

How cutlery is silver-coated by ions.

In this investigation you will find out more about the idea of electrolysis by carrying out your own electro-plating.

What you need

clean iron nail
impure copper strip
supply of very dilute copper sulphate solution
4-12 volt d.c. supply
leads and crocodile clips
beaker

Never use mains electricity in experiments.
Wash your hands after handling chemicals.

What you do

● Set up the electrolysis apparatus as in **D**.
● Check your circuit with your teacher.
● Switch on the electricity.

D

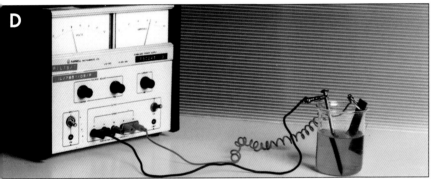

Electrolysis of copper sulphate solution.

1. Here is a list of statements describing your electrolysis experiment. Copy and complete the sentences and write down their correct order.

(a) A .... coloured layer appeared on the anode.
(b) The impure copper electrode went ....
(c) The .... sulphate solution is a light blue colour at the start.
(d) The iron electrode is a .... colour at the start.
(e) The copper electrode was connected to the black, negative electricity terminal. That electrode is called the ....
(f) The impure copper is connected to the .... terminal. It becomes the anode.
(g) Things begin to happen slowly when the .... is switched on.
(h) Pure copper appears on the cathode and the anode gets .... as the copper dissolves.
(i) The impurities in the copper anode do not .... and they sink to the bottom.
(j) The copper in the anode dissolves to form copper ....
(k) The copper ions travel to the .... where they collect as a layer of pure copper metal.

2. Predict what changes you would observe if you increased:
(a) the voltage
(b) the strength of the copper sulphate solution.
If possible, test your prediction by an experiment.

Observations

Record what you see happening in the beaker.

E

Copper-coated leaves.

Now try this . . .

Try electro-plating a dried leaf by coating it with quick-drying glue and then sprinkling it with carbon powder. If you make the leaf the cathode in a beaker of very dilute copper sulphate solution and use a copper strip as the anode, find out what conditions you need to coat the leaf with copper (**E**).

What job does the carbon powder do?

HOW FAST DO REACTIONS GO?

In the Unit "Chemical Changes", you read that making bread using yeast, flour, sugar and so on depends very much on the temperature at which you leave the dough to rise. It also depends on the temperature of the oven in which you bake the bread (**A**). At a low temperature, the bread-making reaction would go much slower than in a warm kitchen. Temperature is a factor that decides how fast a reaction will go.

The students in **B** are observing a reaction being demonstrated by their teacher, Ms. Higgins. She is showing them the following reaction:

What change has happened here?

> **calcium carbonate + hydrochloric acid ⇨ calcium chloride + carbon dioxide gas**

The students can see how fast the reaction is going by measuring the mass of the chemicals left in the flask at regular time intervals. Ms. Higgins asks the students to suggest how they could make the reaction go faster if they were going to demonstrate it.

Ms.Higgins: How can you make the reaction go faster?

Hassan: I could add more acid.

Alex: I could use hot marble lumps

Maria: I could use stronger acid.

Arifa: I could use smaller lumps of marble.

James: I could shake the flask.

Marble chips

Dilute hydrochloric acid.

---

1. You can see the ideas the students have thought of in **B**.
   For each one, predict whether it would make the reaction go faster, slower or make no change.

2. Discuss your answers with your partner and see if you can think of a reason why there would or would not be any change.

In this investigation, you will repeat Ms. Higgins' demonstration and test one of the students' ideas to find out if your prediction is correct.

Wear goggles and protective clothing when handling acid. Wash your hands after handling chemicals.

EXPERIMENT 1
Ms. Higgins' Demonstration.

What you need

large calcium carbonate (marble) lumps
supply of dilute (1M) hydrochloric acid
conical flask
access to an accurate weighing balance
stop clock or digital watch
$100cm^3$ measuring cylinder
filter paper circle

What you do

Read all of this method before you start the experiment.
● Decide what jobs different people in the group are going to do.
● Measure $50cm^3$ of dilute hydrochloric acid into the flask.
● Weigh out 1g of calcium carbonate (marble) lumps on a piece of filter paper.
● Place the flask on the balance and make sure that you can read the scale accurately. Ask for help if you are not sure.
● Add the calcium carbonate quickly to the acid.
● Start the stop watch and write down the mass of the flask plus its contents.
● Record the mass of the flask every 10 seconds for the first 30 seconds and then every 30 seconds afterwards for about 10 minutes.

EXPERIMENT 2
Prediction testing.

What you need

This will depend on your plan. Make a list of apparatus for your teacher to check.

What you do

● Plan the test your group is going to follow.
● Make sure you know what variable you are going to change and what variables you are going to keep the same.

Observations

Record your measurements in a table similar to Experiment 1.

Observations

C

Time after adding calcium carbonate	Mass(g)	Change in mass (g)
0 seconds		
10 seconds		
20 seconds		

3. Use the tables you have filled in to draw line graphs of the experiments. The graphs should have **time** (seconds) along the horizontal axis and **change in mass** (grams) up the vertical axis. You might be able to draw both lines on the same graph.

4. For each idea tested by your group or by your class, infer whether the speed (or **rate**) of reaction has been changed.

5. List the ideas that made the reaction go faster. Which were best?

6. Hassan now says: "If you can make the reaction go faster, you must make more carbon dioxide gas." Do you think Hassan is correct? Discuss this question with your teacher.

Now try this . . .

Limestone is a building material made from calcium carbonate.
In an area where a lot of acid rain falls, at what time of year would you expect the buildings to be attacked most?
How could you stop or slow down the dissolving of the buildings?

GREENHOUSE EARTH

The Earth is a giant spinning ball, moving through space. Although it may seem to be a self-contained spaceship, it is affected by other objects in the **solar system**, especially our nearest **star**, the sun.

Our Earth is unique in the solar system of planets in having an atmosphere containing a mixture of gases that we call "air". Scientists have observed different layers of air in the atmosphere and believe that these layers behave in different ways (**A**).

Where does the atmosphere become space?

Not all of the sun's rays that arrive at the edge of the Earth's atmosphere reach the ground. There are barriers to the sun's radiation that decrease the amount that can be used when it reaches the surface (**B**).

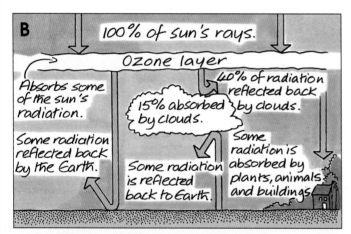

How radiation is reduced before it hits earth.

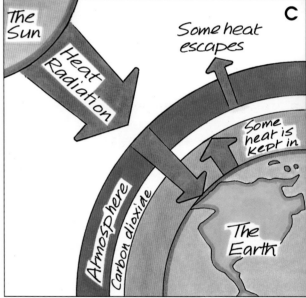

How might the greenhouse effect change life on Earth

1. If there were no clouds, what would happen to the temperature of the Earth?

2. Imagine that the amount of ozone gas in the air decreased. What would happen to the temperature of the Earth?

3. Carbon dioxide gas in the atmosphere lets the sun's radiation through to the surface, but won't let all of it escape if it is reflected back from the Earth (**C**). What would be the effect on the temperature of the Earth if a lot of extra carbon dioxide was made by car exhausts?

4. Green plants, especially trees, use carbon dioxide to make food by a chemical reaction called photosynthesis. They give out oxygen gas in return.
What will be the likely effect on the atmosphere of cutting down large areas of forest?

5. If you were a Government minister responsible for the environment, what steps would you take to improve the quality of the environment?

These activities will help you to understand some more ideas about how the Earth is heated by the sun.

ACTIVITY 1

D shows three of the sun's rays striking the Earth. Measure carefully the distance that the rays **A**, **B** and **C** have travelled to reach the Earth's surface.

D

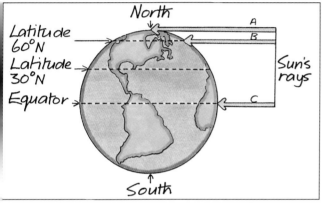

6. What do you infer about the distance that the rays have to travel to reach the equator compared with the North pole?

7. What effect will this have on the amount of heat radiation reaching those places?

ACTIVITY 3

Pictures **F**, **G** and **H** have been taken at places **A**, **B** and **C** - but not in that order!

F

11. Match the photographs with the correct places on the Earth.

ACTIVITY 2

E takes a closer look at the sun's radiation when it reaches the Earth.
Each yellow ray represents the same amount of sunlight.

E

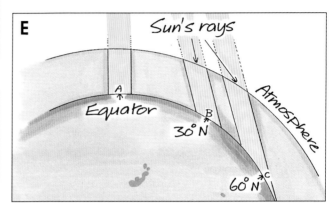

8. Measure the distance on the Earth's surface lit by each of the rays.
   How similar or different are the distances?

9. How will the differences affect the temperatures at **A**, **B** and **C**?

10. Explain how the amount of atmosphere that each of the three sun's rays pass through will affect the temperature at **A**, **B** and **C**.

G

H

MOONS

The name **moon** really refers to a large piece of rock moving around (orbiting) a planet. Our moon is one of many in our solar system (**A**). Pluto also has 1 moon, while Mars has 2. Neptune has 8, Uranus 15, Jupiter 16 (**B**) and Saturn 18. Since many of them are much smaller than our own moon and they are so far away, there is a possibility that more will be discovered.

For many years, people have studied our moon using telescopes. Some people believed that they saw faces in the moon and that they could predict world events by observing changes in their appearance! As telescopes have improved and we have started to send exploring spacecraft to the moon, it has become clear that the marks and wrinkles are mountain ranges and dish shapes called **craters** (**C**).

The first man on the moon.

One of Jupiter's moons.

What might have made this crater?

Jagtar believes that the moon once had volcanoes and the craters are the remains of the volcanoes after they stopped throwing out lava (**D**).

Sally disagrees. She thinks that the craters were made when some huge rocks hurtling through space hit the moon (**E**).

Jagtar imagines volcanoes on the moon.

Sally imagines rocks hitting the moon.

1. Do your observations of the moon support both hypotheses? If not, can you suggest a different explanation for the craters?

2. What evidence would you look for on the moon to support Jagtar's, Sally's or your own ideas?

3. Design simple experiments to see if any of their ideas can be tested using models.
   If possible, try out your tests.
   Do your new observations support any of the hypotheses?

The changing appearance of the moon puzzled people for many years. It was clear that the changes were repeated at regular intervals of about 28 days.

These changes are called the **phases** of the moon (**F-I**). The appearance of the moon at each phase seems to depend on the position of the moon, the Earth and the sun (**J**).

New moon
(solar eclipse).

The four phases of the moon.

7-day moon.

Full moon.

21-day moon.

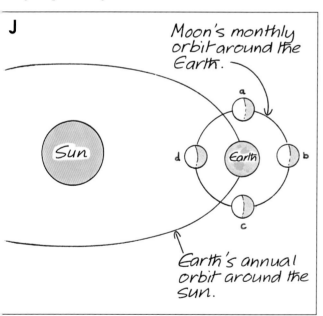

How the phases of the moon are produced.

Producing a "new moon" in a darkened room.

4. Infer which of the positions of the moon in **J** will produce each of the four phases shown in the moon photos (**F-I**).
   If you have difficulty with answering the question, use a torch as the sun's rays and an orange as the moon. You can be the observer on Earth.
   Try to produce the moon's phases, perhaps in a darkened room (**K**).

5. What is the evidence that suggests that the moon keeps the same face towards the Earth at all times?

Now try this . . .

There have been several expeditions to the moon that have brought back samples of moon rock. If you were given some moon rock to examine for signs of life, suggest how you would go about it.
Plan any experiments you think might help.

PLANETS

Our Earth is a **planet**. We belong to a set of planets called the solar system. The nine planets in our solar system follow a path or orbit around a star, the sun. Between the orbits of Mars and Jupiter is a belt of tiny planet fragments called the **asteroids** (**A**).

Our solar system.

PLANETS FACTFILE

MERCURY

- 3000 miles across
- 36 million miles from the sun
- solid surface
- too hot for living things

EARTH

- 8000 miles across
- 93 million miles from the sun
- large area of solid surface
- water present in seas and oceans
- atmosphere supports living things

MARS

- 4000 miles across
- 140 million miles from the sun
- solid rocky surface
- very little atmosphere

VENUS

- 8000 miles across
- 67 million miles from the sun
- solid surface
- poisonous atmosphere

JUPITER

- 87000 miles across
- 480 million miles from the sun
- largely made of very dense gas
- small solid centre of rock and metal
- surrounded by a thin circle of rings containing rocks

SATURN

- 71000 miles across
- 890 million miles from the sun
- made of very dense gas
- dense solid centre of rock and metal
- surrounded by millions of thin rings containing rocks

URANUS

- 29000 miles across
- 1800 million miles from the sun
- made of very dense gas
- dense solid centre of rock and metal
- surrounded by a thin circle of rings containing rocks

NEPTUNE

- 28000 miles across
- 2800 million miles from the sun
- made of very dense gas
- dense solid centre of rock and metal
- no rings detected so far

PLUTO

- 4000 miles across
- 3700 million miles from the sun
- solid surface
- no atmosphere
- too cold for living things

What planets would you visit?

C

PLANET	ORBIT TIME (years)
Mercury	0.2
Venus	0.6
Earth	1.0
Jupiter	11.9
Saturn	29.5
Uranus	84.0
Pluto	248.0

1. Make 2 lists.
   LIST 1: The Solid Planets
   LIST 2: The Gas Planets
   In each list, arrange the planets in order of size, the largest at the top.

2. Choose a planet from each list.
   Using the information available to you, write a postcard to a friend describing what the planet is like (**B**).

Now try this . . .

Table **C** shows how long it takes for seven of the planets to orbit the sun.
Use the PLANETS FACTFILE to draw a line graph of "**time to orbit sun**" against "**millions of miles from sun**". Take care to work out the scales.
From your graph, estimate the orbit times for Mars and Neptune.

On a clear night, you can observe a large number of **stars**, some glowing brightly, others less easy to see. Observations and measurements made by scientists suggest that many of the stars we see are clustered together in groups called **galaxies** (**A**). It is impossible to imagine the distances between us and the other galaxies, even the nearest ones, and scientists can only guess at the number of galaxies in the entire **universe**.

The night sky.

The Earth belongs to a set of planets circling a star, the sun, that is a member of one galaxy called the **Milky Way**. Our sun is a white-hot ball of gas because of the chemical reactions going on inside it (**B**). It is not "burning" in the way that the fuels we use at home do, because it does not use oxygen in its reactions.

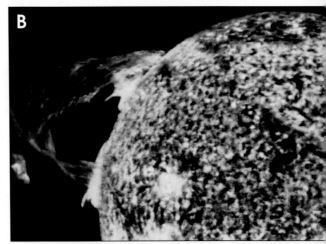

The sun, showing an arch flare.

Why is this plant a good light absorber?

Huge amounts of radiation are sent out from the sun in all directions through space. Some of that radiation is of the type we call **infra-red**, which helps to warm our atmosphere. Another sort of radiation is called **ultra-violet**, which causes skin to darken (tanning) but can also damage animal cells, producing skin cancers. Perhaps the most important sort of radiation produced by the sun is visible light. Sunlight can be absorbed by leaves and other plant parts to help make sugar (**C**) by the reaction called **photosynthesis** (= *making food using light*).

1. List three effects of sunlight on living things.

2. Describe what life would be like without sunlight.

In this activity you will be able to build a simple sundial clock to investigate the movements of the sun and Earth.

Stand up the cut out rectangle.

Hole in paper where rectangle is cut out.

Mark position of shadow

Sun

North

Sheet marked in 5° jumps.

D

Making a sundial.

What you need

stiff white paper or card
protractor
pencil
magnetic compass
scissors

Never look directly at the sun.

What you do

- Take a sheet of paper or card and draw around the protractor.
- Mark the paper with lines at 5° and 10° jumps (**D**).
- Mark the shape of a rectangle and cut around the line.
- On a sunny day, line up the centre line with north, using your compass.
- Stand the rectangle vertically and mark the end of its shadow.
- Repeat the markings every hour, if possible.
- PREDICT some of the marks after you have made two or three.
- If you can, record the shadows at the same time of each day for several weeks or even months.

Observations

Produce a sundial record of the positions of the sun over a long period of time. Write the time and date next to each mark.

3. Describe in words how the pattern of shadow marks changes:
   (a) from hour to hour
   (b) from day to day

4. Mark a sundial card with the hours of the day, so that you can use it as a clock. How accurate is your sundial clock? What are its limits?

5. Since it is dangerous to look at the sun, suggest how scientists can observe the changes that take place on it.

Now try this ...

E shows an extract from a diary. The figures give the times of sunrise and sunset for London at weekly intervals. The times are in Greenwich Mean Time (GMT) with British Summer Time (BST) allowed for from March 31 to October 27.

Draw 2 line graphs of the information, one for sunrise times, the other for sunset times.

Infer a reason for the changes you can see in the graphs.

What effect does British Summer Time have on the graph?

Use your graph to estimate the dates when:
   (a) there is least daylight
   (b) there is most daylight

E

Jan	5	0806	1606	Jul 6	0451	2119
	12	02	16	13	58	14
	19	0756	27	20	0507	07
	26	48	39	27	16	2057
Feb	2	38	51	Aug 3	26	46
	9	27	1704	10	37	34
	16	14	17	17	48	20
	23	00	29	24	59	06
Mar	2	0645	42	31	0611	1951
	9	30	54	Sep 7	22	35
	16	14	1806	14	33	19
	23	0558	18	21	44	03
	30	42	30	28	55	1847
Apr	6	0626	1941	Oct 5	0707	1831
	13	11	53	12	19	15
	20	0556	2005	19	31	00
	27	41	16	26	43	1746
May	4	28	28	Nov 2	0655	1633
	11	16	39	9	0707	21
	18	05	50	16	19	11
	25	0456	59	23	31	02
Jun	1	50	2108	30	42	1556
	8	45	15	Dec 7	51	53
	15	43	19	14	59	52
	22	43	22	21	0804	53
	29	46	22	28	06	58

Sunrise and sunset times.

Tycho BRAHE
1546--1601

HE WENT TO GERMANY FOR MORE TRAINING AND IN 1563 HE OBSERVED A CLOSE APPROACH OF JUPITER AND SATURN, FINDING THAT IT HAPPENED A MONTH AWAY FROM THE TIME PREDICTED IN THE ASTRONOMICAL TABLES. HE BEGAN TO BUY INSTRUMENTS SO HE COULD MAKE OBSERVATIONS FOR THE PREPARATION OF NEW TABLES.

TYCHO BRAHE WAS BORN IN KNUDSTRUP, THEN PART OF DENMARK. HE STUDIED LAW AND PHILOSOPHY AT COPENHAGEN UNIVERSITY, INTENDING TO GO INTO POLITICS. THEN IN 1560 HE SAW A PARTIAL ECLIPSE OF THE SUN AND THIS DECIDED HIM TO CHANGE TO ASTRONOMY. HE WAS TO BECOME THE LAST OF THE GREAT NAKED-EYE ASTRONOMERS.

TYCHO WAS AN EXTREMELY ARROGANT AND QUARRELSOME YOUNG MAN. IN 1565, HE FOUGHT A DUEL WITH SWORDS OVER A POINT IN MATHEMATICS. HIS NOSE WAS CUT OFF AND, FOR THE REST OF HIS LIFE, HE WORE A FALSE NOSE OF METAL.

THE KING OF DENMARK, FREDERICK II, DECIDED TO BECOME TYCHO'S PATRON AND PROVIDED MONEY FOR THE BUILDING OF AN OBSERVATORY ON THE ISLAND OF HVEEN. PART OF THE OBSERVATORY WAS UNDERGROUND TO PROTECT THE INSTRUMENTS FROM THE WIND. THE BUILDING AND INSTRUMENTS COST THE EQUIVALENT OF HALF A MILLION POUNDS IN TODAY'S MONEY.

IN 1572, TYCHO WAS OBSERVING THE SKY AND SAW A NEW STAR – WE NOW KNOW IT WAS A STAR THAT HAD PREVIOUSLY BEEN INVISIBLE TO THE NAKED EYE AND HAD EXPLODED, SO BECOMING MUCH BRIGHTER. HE DESCRIBED THE STAR IN A BOOK 'DE NOVA STELLA', ESTABLISHING THE NAME NOVA FOR 'NEW' STARS.

LEWIS.

IN 1577, A GREAT COMET APPEARED IN THE SKY. *TYCHO'S* OBSERVATIONS SHOWED NOT ONLY THAT THE COMET WAS MUCH FARTHER AWAY THAN THE MOON, UPSETTING THE IDEA OF THE CHANGELESS HEAVENS, BUT THAT ITS ORBIT MUST BE AN ELLIPSE. IF THIS WAS TRUE, THEN IT MUST BE PASSING THROUGH THE VARIOUS PLANETARY SPHERES.

ALTHOUGH *TYCHO* ACCEPTED THAT THIS MUST MEAN THAT THE PLANETARY SPHERES DID NOT EXIST, HE WAS TOO CONSERVATIVE TO ABANDON THE IDEA THAT THE EARTH WAS THE CENTRE OF THE UNIVERSE. INSTEAD HE CAME TO A COMPROMISE – ALL THE PLANETS EXCEPT EARTH REVOLVED AROUND THE SUN. THE SUN, PLUS PLANETS REVOLVED AROUND THE EARTH.

THE ACCURACY OF *TYCHO'S* OBSERVATIONS OF THE SUN AND PLANETS EVENTUALLY MADE A CHANGE IN THE CALENDAR INEVITABLE. IN 1582, *POPE GREGORY* XIII ANNOUNCED THAT THE DATE WOULD HAVE TO BE PUT BACK TEN DAYS TO MAKE THE CALENDAR ACCURATE. THE NEW *GREGORIAN*' CALENDAR WAS FAIRLY QUICKLY ACCEPTED IN CATHOLIC COUNTRIES BUT ONLY SLOWLY IN PROTESTANT ONES. WHEN IT WAS FINALLY ADOPTED IN ENGLAND, IN 1751, ELEVEN DAYS HAD TO BE DROPPED. RIOTS BROKE OUT, WITH PEOPLE CHANTING *'GIVE US BACK OUR ELEVEN DAYS'.*

TYCHO GAVE *KEPLER* HIS OBSERVATIONS AND THE TASK OF PREPARING TABLES OF PLANETARY MOTIONS. WHEN HE DIED IN 1601, *KEPLER* KEPT THE PAPERS AND CONTINUED THE WORK THAT EVENTUALLY LED TO HIS OWN THEORY OF ELLIPTICAL ORBITS. *TYCHO'S* INSTRUMENTS WERE NEVER USED AGAIN AFTER HIS DEATH. WITHIN A DECADE, *GALILEO'S* INVENTION OF THE TELESCOPE MADE THEM OBSOLETE AND THEY WERE FINALLY BURNED DURING THE FIRST YEAR OF THE *THIRTY YEARS WAR.*

IN 1588, *FREDERICK* II DIED AND THE NEW KING, *CHRISTIAN* IV DID NOT HAVE THE SAME PATIENCE WITH *TYCHO'S* TEMPERAMENT. AFTER A FEW YEARS HE STOPPED *TYCHO'S* SUBSIDY AND IN 1597 *TYCHO* LEFT FOR GERMANY AT THE INVITATION OF *EMPEROR RUDOLF* II. HE SETTLED IN *PRAGUE*, TAKING ON AS HIS ASSISTANT THE YOUNG *JOHANN KEPLER.*

UNITS OF LIFE

Cells are often called the "building blocks" of life. All the living things that scientists have observed closely using microscopes are built up from cells - even if it's only one! Pictures **A-D** show four cells, each magnified about 100 times larger than real life. Two of them are from plants, two are from animals. Some of them have been coloured with special stains so that objects inside them can be seen more clearly.

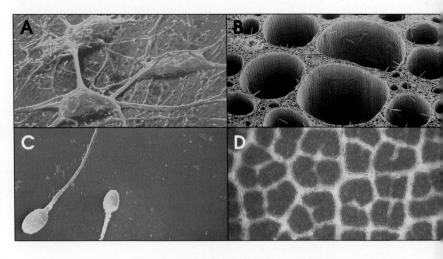

From observations of many plant and animal cells, scientists have identified basic plans for plant and animal cells (**E,F**). They have inferred the jobs, or **functions**, of different parts of the cells from experiments.

1. Pictures **A-D** show cells taken from these parts of living things:

   - plant leaf
   - plant stem
   - animal nerve
   - animal male sex organ

   From what you know about those parts of plants and animals, suggest which photograph goes with each living part.

2. Give a reason for each choice.

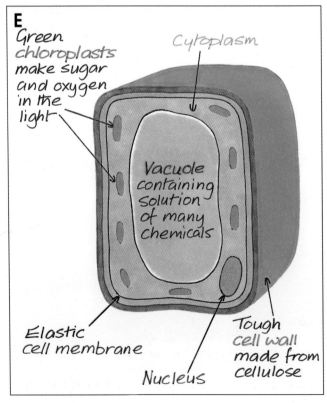

E
Green chloroplasts make sugar and oxygen in the light

Cytoplasm

Vacuole containing solution of many chemicals

Elastic cell membrane

Nucleus

Tough cell wall made from cellulose

A typical plant cell.

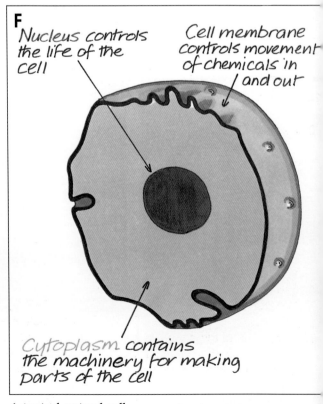

F
Nucleus controls the life of the cell

Cell membrane controls movement of chemicals in and out

Cytoplasm contains the machinery for making parts of the cell

A typical animal cell.

In this experiment you will be able to observe plant cells and compare your observations with the basic plan.

What you need

fresh onion skin
microscope slide
glass cover slip
forceps
iodine solution in dropper bottle
microscope and lamp

 Do not use direct sunlight to observe microscope slides.

What you do

● Use forceps to tear off a thin sheet of fresh onion skin (**G**).
  It should be as thin a sheet of cells as you can tear.
● Place the sheet of cells on the slide and add a drop of iodine solution.
● Carefully lower a cover slip on to the stained cells (**H**).

Observations

Draw the appearance of the cells under different magnifications.

Tearing off onion skin.

Staining cells.

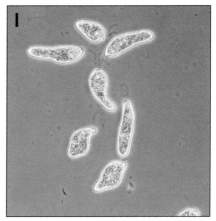

Euglena - plant or animal?

3. Label your drawings with:
  (a) their magnification
  (b) the parts of the cells that you can identify

REMEMBER
Total magnification = eyepiece lens magnification x objective lens magnification.

4. Say what job the iodine does in this experiment.

5. Which part of the basic plant cell plan in E would you NOT expect to observe in an onion cell?
Give a reason for your choice.

Now try this . . .

Copy this table and fill in each column with the parts labelled in the plans **E** and **F**.

PLANT CELLS ONLY	ANIMAL CELLS ONLY	PLANT/ANIMAL CELLS

The organism in **I** is called *Euglena*. It lives in ponds.
  (a) Is *Euglena* a plant or an animal - or neither?
    List any of its features that suggest it is either a plant cell or an animal cell.
  (b) What extra information about *Euglena* would you want to know to support your decision?

FLOWERS AND SEEDS

People often talk about "flowers" when they mean the whole plant. When scientists talk about **flowers**, they mean the part of the plant that is important in reproduction - making **seeds** for a new generation of plants. Flowers come in a huge range of designs (**A**), but they all have the same function to perform. They must help two special cells to meet and join together to make seeds.

What is the function of flowers?

The special cells are called **sex cells** or **gametes**. The male gametes are called **pollen grains** (**B**) and the female gametes are called **ovules** (**C**). Most plants produce pollen grains and ovules, but they are made in different parts of the flower (**D**). The "joining together" of gametes is called **fertilisation**. A fertilised ovule can grow into a seed.

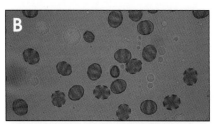

Pollen grains under microscope.

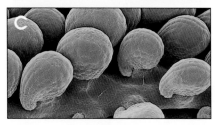

Ovules under microscope.

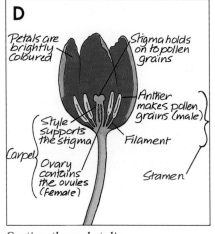

Section through tulip.

Labels in diagram D:
- Petals are brightly coloured
- Stigma holds on to pollen grains
- Anther makes pollen grains (male)
- Style supports the stigma
- Filament
- Carpel
- Ovary contains the ovules (female)
- Stamen

EXPERIMENT 1

What you need

a simple flower, such as a tulip
forceps
safety razor blade or scalpel
transparent sticky tape

Take great CARE when using sharp blades.

What you do

- Using diagram **D** as a guide, take apart the flower you have been given.
- Slice open the **ovary** and take out some ovules.
- Stick some into your book using the sticky tape.
- Dust some pollen on to sticky tape and stick it into your book.

Observations

Make life-size drawings of the separate parts of the flower and label them.

1. Flowers depend on the pollen grains from one plant fertilising the ovules of another. Suggest why it is the pollen that is carried to the ovules and not the other way around.

2. The transport of pollen from one flower to another is called **pollination**. Small animals, such as insects, carry the pollen from the anthers of one flower to the stigma of another.
   Infer what might attract insects to the flowers they pollinate.

In this second experiment, you will observe pollen grains over a period of time and be able to infer some steps in fertilisation.

EXPERIMENT 2

What you need

Do NOT use direct sunlight to observe microscope slides.

flower with anthers
sample of 10% sucrose sugar solution
dropping pipette
microscope cavity slide and cover slip
microscope and lamp
humid container, e.g. sandwich box lined with damp paper towel

What you do

● Smear some pollen on to the cavity slide.
● Add a few drops of sugar solution to the pollen.
● Carefully place a cover slip over the cavity.
● Observe some pollen grains under the microscope.
● Put the slide into a humid container which you can seal (**E**).
● Leave the slides in a warm place for at least one hour.
● Take out your slide and observe the pollen grains again.

E

F

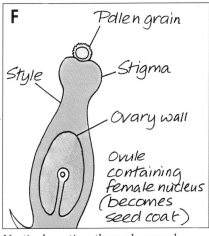

Vertical section through carpel.

Observations

Draw several pollen grains as they look at the beginning and at the end of the experiment.

Now try this . . .

Observe some other flowers and compare the parts you find with the observations you have already made.

REMEMBER! IT IS AGAINST THE LAW TO PICK WILD FLOWERS WITHOUT PERMISSION.

Grass plants have flowers, but they do not need small animals to pollinate them. Find out what grass flowers look like and suggest how they might be pollinated.

3. Describe any changes in the pollen grains you have observed.

4. When Imran looked at his pollen grains after an hour in the sugar solution, he said, "The sugar makes the pollen grains change". Do you agree? Suggest why Imran's inference may not be correct.

5. Why was it important to:
   (a) put the pollen grains in a cavity slide?
   (b) put the slides in a humid container?
   (c) leave the slides for at least one hour?

6. **F** shows what the female part of a flower looks like when it is sliced down the middle. Suggest how the changes you have observed in the pollen grains might help fertilisation of the ovule to happen.

7. What must happen to the female part of the flower so that the seeds can escape?

NEW PLANTS

When seeds have been made, it is important to transport them away from the parent plant to give the new plants a good chance of survival. There are several ways in which plants can **disperse** their seeds to fresh growing places. **A** and **C** show some of the seeds that drop from a sycamore tree (**B**).

1. Why is it important for seeds to get as far from the parent plant as possible?

2. What should the conditions be like for a plant to disperse its seeds?

3. How are sycamore seeds designed to give them the best chance of getting away from the tree?

4. Plan an experiment to investigate the hypothesis that "sycamore seeds with longer wings go further than seeds with shorter wings".

What is the function of the papery "wings"?

Many plants disperse their seeds with the help of containers of various sorts. These containers can be **fruits**, **berries** or **pods**. Some examples of these are described in **D**.

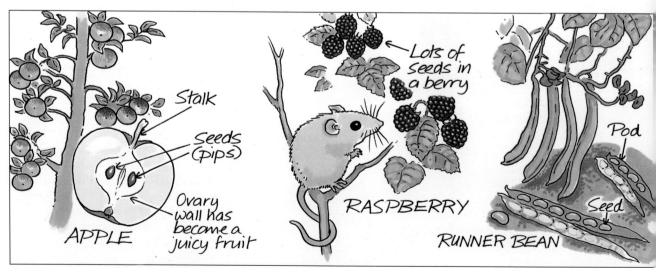

How are these seeds dispersed?

5. For each example in **D**, suggest how the seeds can be dispersed to new growing places.

6. Choose a piece of fruit not shown in **D**. Cut it open and look for the seeds. Infer how the seeds might be dispersed from the fruit.

When conditions are right seeds will begin to grow or germinate. In this investigation, you will grow some young plants and collect information for a discussion on germination.

Dave, Joe and Audrey were having a chat about germinating runner beans (**E**).

E

I always sandwich the seeds between layers of compost and put them in the airing cupboard. It's the warmth they like.

The compost is the best thing, but you shouldn't put them in the dark. They won't know which way up to grow.

I thought seeds grew at any temperature.

What you need

a selection of seeds
containers for germinating seeds and growing seedlings
seed compost
blotting paper

What you do

● Plan a set of experiments to investigate the ideas of Joe, Dave and Audrey.
● You may need to ask for other apparatus to help you. Make a list.
● Plant your seeds and observe the changes that take place over several lessons.
● Measure the size of as many of the small plants (**seedlings**) as you can (**F**).

F

7. Draw line graphs to show how the size of an average plant from each of your experiments changed during the investigation.

8. From the observations you have made, say how strongly you agree with the ideas of Joe, Dave and Audrey.

9. What variables did you control to make your investigation a fair test of their ideas?

10. Dave was looking through his gardening book and saw a diagram of the inside of a broad bean seed (**G**).
"Do all seeds have the same parts?" he wondered.
Using your own seeds, try to find an answer to Dave's question.

Observations

Record your observations and measurements in a "seedling diary".
This should include drawings and tables.

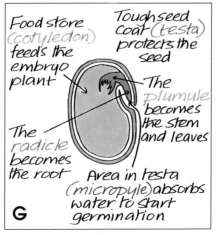

Food store (cotyledon) feeds the embryo plant

Tough seed coat (testa) protects the seed

The plumule becomes the stem and leaves

The radicle becomes the root

Area in testa (micropyle) absorbs water to start germination

G

Section through broad bean.

Now try this . . .

Find out the science meanings of these words and link the meanings to how some plants make new ones.
  (a) bulb
  (b) runner
  (c) tuber

Give an example of a plant that can reproduce in each way.

FOOD FROM THE SUN

Scientists who are growing the tomato plants in the greenhouse shown in **A** are confident that they will grow better than the tomato plants grown outdoors in **B**.

1. Suggest why the scientists can be sure that the tomato plants growing in the greenhouse will do better than the outdoor ones.

2. How could you tell that tomato plants grow better in the greenhouse than outdoors?

Tomatoes grown under glass.

Tomatoes grown in open air.

There is a chemical reaction that goes on inside the green parts of plants called **photosynthesis**. As the name suggests, it is a synthesis or combination reaction and it describes the joining together of two simple compounds, **water** and **carbon dioxide**, to make **sugar** and a vital extra product, **oxygen gas**:

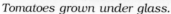

| water + carbon dioxide gas ⇨ glucose sugar + oxygen gas |

The energy to make the reaction happen comes from light - the "photo" part of photosynthesis. Experiments by plant scientists have shown that packets of a green compound called **chlorophyll** can absorb light and use it to synthesise sugar and oxygen from water and carbon dioxide. These packets of chlorophyll are called **chloroplasts** and it is the chloroplasts that make a plant look green (**C**).

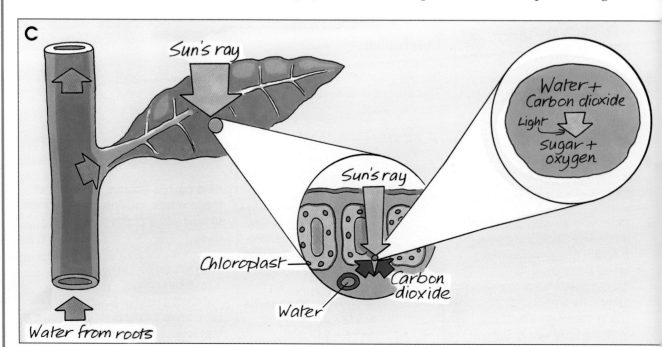

How are plants designed to be "sugar factories"?

Scientists believe that the glucose sugar is stored inside plants as a bigger compound called starch. In this investigation, you will look for starch as evidence for photosynthesis.

What you need

healthy growing plant (geranium or tomato)
black plastic sheet or card
iodine solution
Bunsen burner and heat-proof mat
tripod and gauze
beaker of water
warm ethanol
tweezers
white tile
Resource Book Sheet "Starch Testing"

> ⚠ **Wear safety goggles when heating. Ethanol is flammable - keep away from flames.**

What you do

- Completely cover one of the leaves of the plant with black plastic or card (**D**).
- Leave the plant to grow for 2 or 3 days.
- Remove the cover from the leaf and cut off a piece of the leaf.
- At the same time, cut off a piece of a leaf that was not covered.
- Carry out a search for starch on each piece of leaf by following the test described on the Teachers' Resource Sheet "Starch Testing".

D

Observations

Draw the appearance of each piece of leaf after you have tested them.

E

A positive starch test.

F

This plant has variegated leaves.

3. From your observations, infer which piece of leaf contains most starch. Say how you can tell.

4. Infer which of the leaves carried out most photosynthesis.

5. What difference would you expect to see in the starch test if the leaf had been covered for twice as long?

6. **F** shows leaves described as **variegated**.
   (a) How are they different from the usual sort of non-variegated leaf?
   (b) Draw how you think one of them would look if you carried out a starch test on it.

Now try this . . .

The chemical called potassium hydroxide absorbs carbon dioxide gas. Design an experiment to investigate the hypothesis that "plants do not grow if they cannot get carbon dioxide".

Make clear what variables you are changing and which you are keeping the same.

Predict what observations you would expect to make. If possible, carry out your test and write a report of your investigation.

SIMILAR OR DIFFERENT?

When a new human baby is born, everybody in the family wants to know who the baby most looks like. Has Ben got Grandpa's nose? Has Dipti got her aunty's smile? We expect members of the same family to have similar features. Scientists say that the baby has **inherited** the features from its parents.

It isn't just in humans that features or **characteristics** are inherited. The offspring of all animals resemble the animals that mated to produce them (**A**).

What features has this lamb inherited?

Puppies from the same litter.

The same is true of plants. Seeds are produced as a result of one plant being fertilised, usually by the pollen of another. The plants that germinate and grow from those seeds show certain features that are special to their "parents" or even "grandparents". They will also show small differences or **variations** that had not been seen before. The members of any family will have variations that allow them to be separated (**B**). Only identical brothers or sisters have such little variation that they look so similar.

1. List 10 features that you think can be passed on from human parents to their offspring.

2. Try to produce another list of features that you think might NOT be decided by inheritance.

3. Look at any plants that you have available. Can you list 10 plant features that might be passed on from one generation to the next?

The passing on and mixing of features can only happen at fertilisation, when the male and female sex cells or **gametes** (sperms and eggs in animals, pollen grains and ovules in plants) join together. Experiments have shown that the nucleus of each sort of gamete contains a special chemical code that decides what the offspring will look like. Small packets of this chemical are called **genes**. When the genes from each gamete mix together, they make a new combination that may develop into an individual with new variations (**C**).

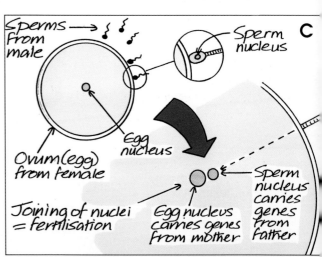

Variations that are decided only by the genes a living thing inherits are called **discontinuous**. An example would be the colour of our eyes or our blood group.

Variations that are decided mostly by how healthy the offspring is and how well it grows are called **continuous** variations. An example would be the length of our feet or the height of a tree.

These activities look at continuous and discontinuous variation in plants and animals. They ask you to think about the differences between them.

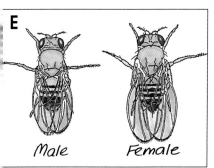

E

Male and female fruit flies.

4. Draw a histogram graph of your seed measurements. This should show how many of the seeds fall into different size ranges. An example of a histogram is shown in **F**.

5. Is "seed size" an example of continuous or discontinuous variation? Give a reason for your answer.

6. Describe the different varieties of fruit fly you found on the Resource sheet.

7. Is "fruit fly variety" an example of continuous or discontinuous variation? Give a reason for your answer.

8. What observation can you make that suggests that the height of people has something to do with genes?

What you need

sample of about 50 large seeds (peas, beans, sunflower, or wheat grains)
30-50cm. ruler
Resource Book sheet "Fruit Flies"
graph paper

What you do

SEEDS
● Draw a table similar to the one in **D**.
● Measure the length of each of the seeds given to you, to the nearest millimetre.
● Record the measurements in your table.

FLIES
● Fruit flies are quite common in the summer. They feed on the sugary juices of ripe fruit.
● Collect the Resource Book sheet "Fruit Flies" from your teacher.
● **E** shows the appearance of a male and a female fruit fly of the same variety.
● How many varieties can you find among the flies shown on the sheet?

Observations

D

> Write down your measurements in a table like this one.

Length of seeds (mm)	Tally	Total
5 – 7	I I I I	4
8 – 10	I I I	3
11 – 13	I I I I I	6

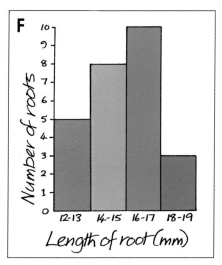

F

Length of root (mm)

(y-axis: Number of roots, 0–10; x-axis: 12-13, 14-15, 16-17, 18-19)

Now try this . . .

Which of the following features are examples of continuous variation and which are examples of discontinuous variation?
Give a reason for your choice in each case.

(a) sex
(b) muscle strength
(c) shoe size
(d) hair colour
(e) foot length
(f) number of roots
(g) breed of dog

CELL DIVISION

You have seen that a gene is the basic unit of inheritance. Each gene controls one characteristic, such as hair or flower colour, although most features are decided by several genes, not just one. There are so many characteristics to build into even a simple organism that many thousands of genes are needed. Genes do not float around free inside a cell, they are joined together into microscopic threads called **chromosomes** (A). The chromosomes are held together inside the nucleus of a cell until it is ready to divide.

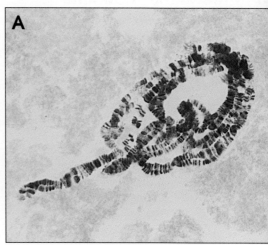

A chromosome stained to show genes.

Nearly all the cells of a human contain the same number of chromosomes, 46, while the cells of a very different organism will have a different number, such as 8 in a fruit fly (B).

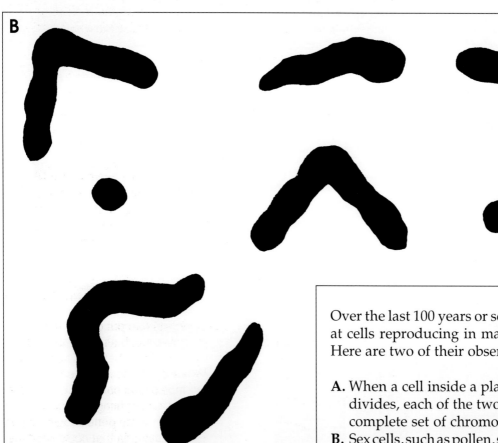

Chromosomes of a fruit fly.

Over the last 100 years or so, scientists have looked at cells reproducing in many plants and animals. Here are two of their observations:

A. When a cell inside a plant or an animal's body divides, each of the two new cells contains a complete set of chromosomes.

B. Sex cells, such as pollen, sperms, eggs and ovules contain half the number of chromosomes found in cells from the rest of the organism.

1. Why do you think that chromosomes were not discovered until about 100 years ago?

2. What must happen to the chromosomes when a cell divides to explain observation **A**?

3. Explain why observation **B** is important.

4. Make another observation of the chromosomes in picture **B** that the scientists should also have noticed.

This activity will help you to understand how the chromosomes behave when a cell divides. It asks you to make a "flick book" showing scenes from the life of a cell!

Observations

Flick your book to see a cell divide!

What you need

Resource Book sheet "Cell Division"
card
glue
scissors
sticky tape
coloured pencils

What you do

- Collect the Resource Book sheet "Cell Division" from your teacher.
- The rectangle shapes contain diagrams of chromosomes at different stages of cell division.
- Decide on the most likely order of the pictures, starting with Number 1 and ending with Number 20.
- Fill in two of the pictures for yourself to complete the series.
- You might like to colour in the chromosomes.
- Glue the sheet to a sheet of card and then cut out the rectangles.
- Arrange the 20 cards into a pile, with card 1 at the top and card 20 at the bottom.
- Staple the edges of the cards together to make a "flick book" (**C**).

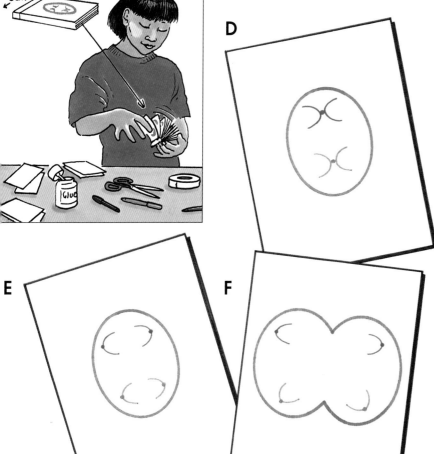

5. Explain the difference between a gene and a chromosome.

6. Describe in words what is happening in these three cards taken from the flick book (**D**, **E**, **F**).

Now try this . . .

Scientists can put genes into microbes, such as bacteria, that were not there before. These new genes direct the microbes to produce useful compounds that are important in fighting diseases. However, some genes could be put into the bacteria that could produce dangerous compounds if they escaped from the lab.

Write a letter to a friend explaining why you think it is a good or a bad idea to experiment with genes and bacteria.

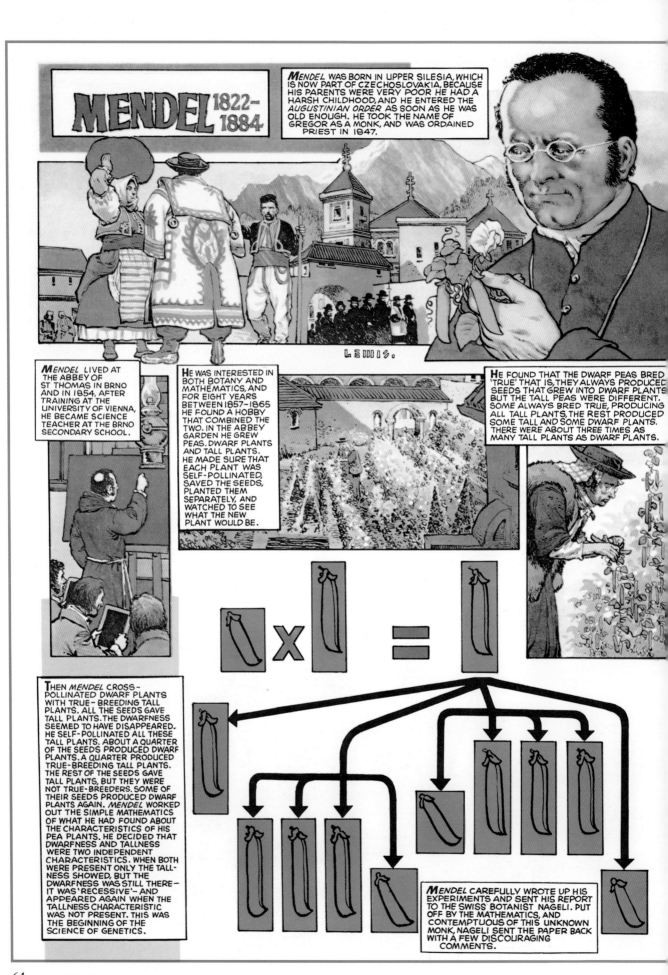

MENDEL 1822–1884

MENDEL WAS BORN IN UPPER SILESIA, WHICH IS NOW PART OF CZECHOSLOVAKIA, BECAUSE HIS PARENTS WERE VERY POOR HE HAD A HARSH CHILDHOOD, AND HE ENTERED THE *AUGUSTINIAN ORDER* AS SOON AS HE WAS OLD ENOUGH. HE TOOK THE NAME OF GREGOR AS A MONK, AND WAS ORDAINED PRIEST IN 1847.

MENDEL LIVED AT THE ABBEY OF ST THOMAS IN BRNO AND IN 1854, AFTER TRAINING AT THE UNIVERSITY OF VIENNA, HE BECAME SCIENCE TEACHER AT THE BRNO SECONDARY SCHOOL.

HE WAS INTERESTED IN BOTH BOTANY AND MATHEMATICS, AND FOR EIGHT YEARS BETWEEN 1857–1865 HE FOUND A HOBBY THAT COMBINED THE TWO. IN THE ABBEY GARDEN HE GREW PEAS. DWARF PLANTS AND TALL PLANTS. HE MADE SURE THAT EACH PLANT WAS SELF-POLLINATED, SAVED THE SEEDS, PLANTED THEM SEPARATELY, AND WATCHED TO SEE WHAT THE NEW PLANT WOULD BE.

HE FOUND THAT THE DWARF PEAS BRED 'TRUE' THAT IS, THEY ALWAYS PRODUCED SEEDS THAT GREW INTO DWARF PLANTS BUT THE TALL PEAS WERE DIFFERENT. SOME ALWAYS BRED TRUE, PRODUCING ALL TALL PLANTS, THE REST PRODUCED SOME TALL AND SOME DWARF PLANTS. THERE WERE ABOUT THREE TIMES AS MANY TALL PLANTS AS DWARF PLANTS.

THEN *MENDEL* CROSS-POLLINATED DWARF PLANTS WITH TRUE-BREEDING TALL PLANTS. ALL THE SEEDS GAVE TALL PLANTS. THE DWARFNESS SEEMED TO HAVE DISAPPEARED. HE SELF-POLLINATED ALL THESE TALL PLANTS. ABOUT A QUARTER OF THE SEEDS PRODUCED DWARF PLANTS. A QUARTER PRODUCED TRUE-BREEDING TALL PLANTS. THE REST OF THE SEEDS GAVE TALL PLANTS, BUT THEY WERE NOT TRUE-BREEDERS. SOME OF THEIR SEEDS PRODUCED DWARF PLANTS AGAIN. *MENDEL* WORKED OUT THE SIMPLE MATHEMATICS OF WHAT HE HAD FOUND ABOUT THE CHARACTERISTICS OF HIS PEA PLANTS. HE DECIDED THAT DWARFNESS AND TALLNESS WERE TWO INDEPENDENT CHARACTERISTICS. WHEN BOTH WERE PRESENT ONLY THE TALL-NESS SHOWED, BUT THE DWARFNESS WAS STILL THERE – IT WAS 'RECESSIVE'– AND APPEARED AGAIN WHEN THE TALLNESS CHARACTERISTIC WAS NOT PRESENT. THIS WAS THE BEGINNING OF THE SCIENCE OF GENETICS.

MENDEL CAREFULLY WROTE UP HIS EXPERIMENTS AND SENT HIS REPORT TO THE SWISS BOTANIST NAGELI. PUT OFF BY THE MATHEMATICS, AND CONTEMPTUOUS OF THIS UNKNOWN MONK, NAGELI SENT THE PAPER BACK WITH A FEW DISCOURAGING COMMENTS.

MENDEL PUBLISHED HIS REPORT, BUT ONLY IN THE *TRANSACTIONS* OF THE LOCAL NATURAL HISTORY SOCIETY. AT THE SAME TIME THE PRUSSIANS, UNDER THE LEADERSHIP OF BISMARK, HAD CONQUERED AUSTRIA AND OCCUPIED GREAT AREAS OF COUNTRY, INCLUDING BRNO. MENDEL'S PAPER WAS FORGOTTEN.

AFTER THIS, *MENDEL* DID NO MORE RESEARCH. E WAS DISCOURAGED BY NAGELI'S REBUFF, ND FULLY OCCUPIED IN THE RUNNING OF THE MONASTERY, WHERE HE WAS NOW ABBOT. ESIDES, HE WAS GROWING *TOO PLUMP* TO ARRY ON WITH HIS GARDENING! *MENDEL* DIED IN 1884, NEVER KNOWING HAT ONE DAY HE WOULD BE FAMOUS. IN 1900 HREE SEPARATE BOTANISTS, ONE DUTCH, ONE ERMAN AND ONE AUSTRIAN, EACH WORKED UT INHERITANCE THEORY—AND THEN ISCOVERED THAT *MENDEL* HAD DONE IT ALL BEFORE THEM.

UT OUT 50 SQUARES OF BLUE CARD AND O SQUARES OF BROWN CARD. THESE EPRESENT THE 'GENES' FOR BLUE AND ROWN EYES. THE BLUE GENES ARE *ECESSIVE*. FIRST LAY DOWN TWO BLUE ARDS FOR A PURE BLUE-EYED MOTHER, ND TWO BROWN CARDS FOR A PURE ROWN-EYED FATHER. WHEN THESE ARE IXED THE CHILDREN WILL EACH HAVE NE BLUE CARD AND ONE BROWN CARD. HESE CHILDREN ARE ALL BROWN-EYED. O OUR 50 BLUE CARDS AND 50 BROWN ARDS REPRESENT 50 BROWN-EYED HILDREN, WHO ARE GOING TO MARRY ND HAVE CHILDREN OF THEIR OWN. HUFFLE ALL THE CARDS TOGETHER ND DEAL THEM OUT IN PAIRS. BROWN-BLUE PAIR REPRESENTS A ROWN-EYED CHILD, AND A BROWN-ROWN PAIR REPRESENTS A BROWN-YED CHILD, BUT A BLUE-BLUE PAIR EPRESENT A BLUE-EYED CHILD.

YOU WILL FIND THAT YOU GET ABOUT THREE TIMES AS MANY BROWN-EYED CHILDREN AS BLUE-EYED CHILDREN. WE BEGAN WITH GRANDPARENTS WITH PURE BLUE AND BROWN EYES. THEIR CHILDREN ALL HAD BROWN EYES. BUT A QUARTER OF THE GRANDCHILDREN HAD BLUE EYES.

A CLOSER LOOK AT GENES

All of the cells in your body, except for your sperm cells or your egg cells, contain 46 chromosomes. When body cells are about to divide, you can see that there are 23 pairs of similar-looking chromosomes. When the sex cells are made, only one member of each of those pairs ends up in a sperm or an egg. It is important for this to happen so that a full set of 46 chromosomes is created when fertilisation happens (**A**).

A

Mother
46 chromosomes in every body cell → Egg – 23 chromosomes

Father
46 chromosomes in every body cell → Sperm 23 chromosomes

Fertilised egg 46 chromosomes

.... divides many times ....

Baby
46 chromosomes in every body cell

The new baby inherits **two** matching sets of 23 chromosomes, one set from its mother and one from its father. Both sets of chromosomes carry genes that have a say in deciding what the baby will look like. This means that you have **two** genes for all of your characteristics. So which set of genes will have the biggest say?

If each parent passes on a gene for brown eyes to a child, then the child will have brown eyes, as you would expect. But the genes inherited from each parent may be different. In that case, the genes that are "stronger" or more **dominant** decide what happens. The "brown eye gene" is stronger than the "blue eye gene", which scientists call the **recessive** gene. If one parent passes on a brown eye gene and the other passes on a blue eye gene, the child will grow up to have brown eyes (**B**).

When scientists who study genes are writing about dominant genes, they give them a single capital letter. The recessive gene for the same characteristic is given the small version of that letter.

So, the brown eye gene is **"B"** and the blue eye gene is **"b"**.
A child who has inherited one of each sort of gene has the **gene type "Bb"**.
It will have brown eyes because **"B"** is dominant over **"b"**.
A child who inherits a **"b"** blue eye gene from both parents will have blue eyes and a gene type called **"bb"**.

B

Father has brown eyes

Mother has blue eyes

Sperm carrying dominant 'brown eyes' gene

Egg carrying recessive 'blue eyes' gene

Baby has brown eyes

Eye colour inheritance.

C

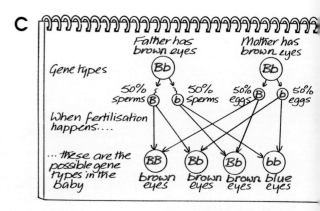

Father has brown eyes **Bb**

Mother has brown eyes **Bb**

Gene types

50% sperms **B**   50% **b** sperms   50% **B** eggs   50% **b** eggs

When fertilisation happens....

... these are the possible gene types in the baby

BB brown eyes   **Bb** brown eyes   **Bb** brown eyes   **bb** blue eyes

C shows a page from a gene scientist's notebook. She is explaining to a couple how their genes can mix when fertilisation happens. Both of them have brown eyes, but they have the gene type **"Bb"**.

66

This activity will help you to understand how you can predict the likely eye colour of children if you know the gene types of their parents.

What you need

50 small squares of brown card. These all represent dominant *"B"* brown eye genes.
50 small squares of blue card. These all represent recessive *"b"* blue eye genes.

What you do

- Take 25 brown and 25 blue cards.
- Shuffle them together and place them as a stack on the table.
- Each of these cards represents the eye gene of a sperm from a *"Bb"* father.
- Shuffle the rest of the cards and stack them as before.
- Each of those cards can represent the eye gene of an egg from a *"Bb"* mother.
- Take the top card from each stack and lay them on the table.
- These two cards represent the gene type of a possible baby.
  Can you predict how many of each gene type you will get?
  Can you predict how many possible children will have blue or brown eyes?
- Carry on "fertilising" an egg card with a sperm card until they are all paired off.

1. How well did you predict the results of the card experiment? If it was difficult, say what the problem was.

2. What fraction of the possible children have brown eyes? What fraction have blue eyes?

3. If you repeated the experiment, would you expect the same results? Give a reason for your answer.

4. Mr. Jones has brown eyes and a gene type of *"BB"*. Mrs. Jones also has brown eyes but her gene type is *"Bb"*. The gene scientist (**E**) says that she is certain that all of their possible children will have brown eyes.
How can she be so sure?

Observations

D

> Copy this table into your book.

Gene type	Eye colour	Tally, e.g. ШТ	Total

> How many 'babies' of each gene type did you count?

E

> This shows that your child is certain to have brown eyes!

Now try this . . .

A couple have gene types *"Bb"* and *"bb"*.
Use your cards to predict the fraction of their possible children that have:
  (a) each possible gene type
  (b) blue eyes

What advice could a gene scientist give to this couple?

INHERITED DISEASE

Our genes decide how the complicated chemicals that make up our bodies are built and work together properly. No two people have exactly the same set of chemicals. There are very small variations that are produced by slight differences in the genes. The effects of some variations in our chemicals are easy to see, like the colour of our eyes or the curliness of our hair. Other variations, like our blood group, are more difficult to notice. Scientists have to do tests on the blood to detect them (**A**). Understanding how small changes in our genes affects our lives is an important aim for gene scientists.

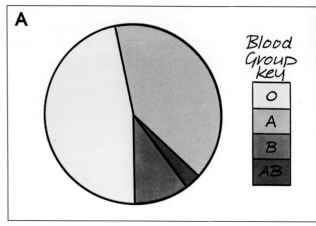

Blood groups in the UK

This girl has cystic fibrosis. She needs regular exercise and treatment to keep her lungs working properly.

Most of us are fit and healthy because all of the chemicals in our bodies are working well together. However, this is not always true. Sometimes, a faulty gene is inherited that produces a slightly different chemical from the one in a healthy body. This small variation can be enough to upset the proper working of the body in an important way. Diseases caused in this way are called **inherited diseases**, such as **cystic fibrosis** (**B**, **C**), and **sickle cell anaemia**.

It is possible for someone to inherit the faulty gene, but not show the full disease. They are called **carriers** of the faulty gene. Carriers do not show the disease because they also inherit a healthy gene for that characteristic from the other parent. Healthy genes are usually dominant over the faulty genes. It is only when two carriers of the same faulty gene have a child that it is possible to see the full disease symptoms (**D**).

How can two carriers of a faulty gene produce an affected child?

1. Which blood group is the most common in the United Kingdom?

2. What is an inherited disease?

3. Name one inherited disease and say how it can affect people.

4. How can someone be a carrier of an inherited disease?

Some scientists who study genes are able to give advice to people about the chances of having children with certain inherited diseases. Read this information about sickle-cell anaemia and then try to give advice to the people who ask for help.

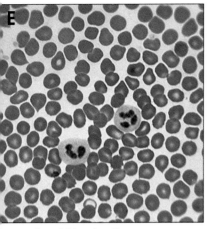

Normal red blood cells.

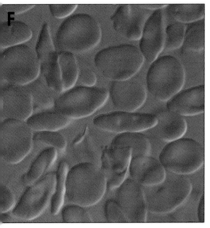

Sickled red blood cells.

SICKLE CELL ANAEMIA FACTFILE

❑ This disease of the blood is common in Africa and among people from African families.
❑ The red blood cells of people suffering from sickle cell anaemia are different from the red cells in healthy people.
❑ The faulty red cells change their shape and block the narrow blood vessels called capillaries.
❑ If capillaries are blocked, blood cannot pass through.
❑ Great pain is caused and important organs of the body can be damaged.
❑ The disease only develops if a child inherits two recessive genes (gene type "*ss*").
❑ Scientists can do a blood test to see who has the gene type "*ss*".

5. Imagine that you are giving advice to couples about sickle cell disease and the chances of having affected children.
   (a) Explain in your own words what the disease is and how it is caused.
   (b) You carry out tests on two couples, **A** and **B**.
   In couple **A**, the husband is not a carrier. He has the gene type "*SS*".
   His wife is a carrier. What is the chance of any of their children suffering from the disease?
   (c) Both the husband and the wife in couple **B** are carriers. What is the chance of any of their children suffering from the disease?

G

Why is it important to have magazines such as this one?

ANIMAL REPRODUCTION

All animals, like other living things, must reproduce. Some simple animals do not have **male** and **female** types and they can reproduce **asexually**, by splitting in half. An example is the single cell pond organism, *Paramecium* (**A**).

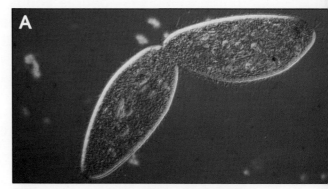

A

In most animals there are two sexes, male and female. The male and female parents have special parts of their bodies where sex cells (or gametes) are made. These parts are called **sex organs**.

How is this pond organism reproducing?

> The male sex organs are called **testes**. They make the sex cells called **sperms**.
> The female sex organs are called **ovaries**. They make the sex cells called **eggs**.

The aim of **sexual reproduction** is to help the gametes to meet during fertilisation. A fertilised egg can then divide over and over again to make a ball of cells or **embryo** (**B**). By making more and more cells, the embryo grows into a new animal.

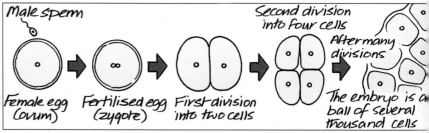

B

Male sperm

Female egg (ovum) → Fertilised egg (zygote) → First division into two cells → Second division into four cells → After many divisions → The embryo is a ball of several thousand cells

From egg to embryo.

Different animals use different methods to help the sperm and egg to meet. In some, such as fish and amphibians, the sperms and eggs join together outside the body of the female animal. Scientists call this **external fertilisation** (**C**). In other animals, including reptiles, birds and mammals, the sperm meets the egg inside the body of the female. This is called **internal fertilisation** (**D**).

C

What is wasteful about external fertilisation?

D

Why is internal fertilisation better for birds?

1. Give 2 reasons why it might be useful to *Paramecium* to be able to reproduce by dividing into two.

2. Suggest how sperm cells are well designed to fertilise eggs.

3. Explain what is meant by the word "embryo".

4. Describe the main difference between internal and external fertilisation.

5. Say what sort of fertilisation you would expect these animals to have, internal or external?
   (a) crow   (b) frog   (c) kangaroo
   (d) snake   (e) trout

The main changes during the life of an animal are part of its life cycle. In this activity you are asked to infer the life cycles of three animals, a fish, a frog and a chicken.

What you need

Resource Book sheet "Life Cycles"
scissors
glue

What you do

- Collect the Resource Book sheet "Life Cycles" from your teacher.
- Cut out the 18 pictures.
- Put them into three groups of six pictures for a fish, a frog and a chicken.
- Arrange each group in a circle and stick them in your book.
- Draw arrows between each picture to complete the Life Cycles for the fish, the frog and the chicken.

Development of a fish.

6. Label the correct cards in each cycle with these words:
  - "adults mating"
  - "sperms fertilising eggs"
  - "embryo growing"
  - "birth"
  - "young animal"
  - "growing into an adult"

7. Give one important difference between the eggs of a frog and the eggs of a bird.

8. All embryos need food to grow.
Explain where the food comes from for the bird embryo.

Development of a frog.

Now try this . . .

Most animals cannot reproduce all year round. They have a mating season when the female's eggs can be fertilised. Scientists believe that most female animals make chemical scents during the mating season to attract males.

Plan an experiment to find out if butterflies make these chemical scents.

Describe other ways in which male and female animals try to attract each other.

Development of a chicken.

THE HUMAN LIFE CYCLE

Humans do not start to produce sex cells, sperms and eggs, until they reach a time of life called **puberty**. Boys usually reach puberty between 12 and 16, while girls might enter puberty rather earlier, between 10 and 14.

Puberty is a time of considerable change in human bodies. Diagrams **A** and **B** show the male and female sex organs at the end of puberty. These changes are summarised in **C**.

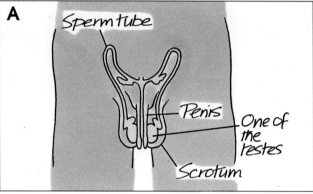

The male reproductive system.

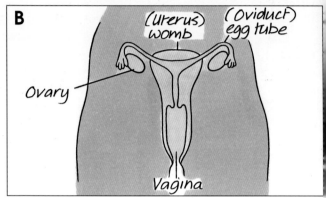

The female reproductive system.

C

FEMALE	MALE
Hair grows under arms.	Hair grows on face.
Breasts develop.	Voice becomes deeper.
Hips become wider.	Hair grows under arms.
Pubic hair grows.	Pubic hair grows.
	Penis grows.

Table of changes during puberty.

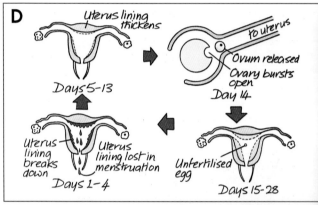

The menstrual cycle.

One of the many partly formed eggs in a woman's ovaries is set free each month. This release of a mature egg is called **ovulation**. The egg may be fertilised if it meets sperms as it travels along the egg tube (**oviduct**). During the journey, the lining of the womb or **uterus** is growing in readiness for a fertilised egg, if one arrives. If the egg is not fertilised then the new uterus lining is not needed. It breaks down and passes out of the vagina together with some blood and the unfertilised egg. This bleeding is called a **period**. A woman will wear an absorbent towel at that time. The beginning of a period marks the start of the next monthly or **menstrual cycle** (**D**).

1. Collect a copy of the Resource Book sheet "Sex Organs". Stick it into your book and label the following parts:
   (a) penis
   (b) ovary
   (c) womb (uterus)
   (d) testes
   (e) scrotum
   (f) vagina
   (g) egg tube (oviduct)
   (h) sperm tube

2. On what day in the menstrual cycle does ovulation usually happen?

3. If a woman releases eggs from the age of 15 until she is 45, how many eggs will she have produced?

Sexual Intercourse

When a couple make love, or have **sexual intercourse**, the male **penis** becomes stiff and erect. It fits into the **vagina** and the man and woman move together. After a short time, the man may become so excited that there is a sudden release of liquid from his penis. This liquid is called **semen** and it contains millions of sperms. The sperms swim through the uterus and into the oviducts. If a sperm fertilises an egg, the fertilised egg, called a **zygote**, is carried into the uterus where it may become fixed in the lining and start to grow into a baby (**E**).

Pregnancy

Over the next 38 weeks or so, the time when the woman is called **pregnant**, the unborn child or embryo grows inside the mother's uterus. Starting off as a ball of cells, the embryo begins to take shape. It is surrounded by a cushion of warm liquid that protects it from damage while it develops. The embryo receives food and oxygen through a flexible cord joined to a part of the mother's uterus called the **placenta** (**F**).

G tells you what changes happen to the embryo as it grows into a mature unborn baby or **foetus**.

How many cells are in this embryo?

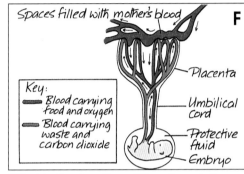

What is the function of the placenta?

4. What protects the baby when it is in the uterus?

5. Why is the placenta important to the embryo?

6. **H** shows how the mass of the embryo changes during pregnancy.
   Estimate:
   (a) the mass of the embryo in week 10.
   (b) the mass of the embryo in week 35.
   (c) the week in which the embryo weighs 1.0kg.
   (d) the weeks between which growth is fastest.

7. How does a new-born baby get food?

8. If a baby is born too early, it is kept warm in an incubator. Why?

G

Week	What happens to baby?
0	Sperm fertilises egg in oviduct.
1	Embryo buries itself in wall of uterus.
2	Embryo's arms and legs appear as tiny bumps. Its eyes begin to develop.
6	Embryo starts to look human. Ears, hands and feet appear. Heart starts beating.
10	Fingers and toes start to grow and it can move its arms and legs.
14	Doctors can tell if foetus is a boy or a girl.
18	Foetus has hair. Mother can feel its kicks!
24	The eyes of the foetus open. If the baby was born now, it could survive if given special care.
34	Foetus has put on a lot of weight, mostly fat to keep newborn baby warm.
38	Mother's uterus squeezes and baby is born.

Now try this . . .

Find out the scientific meaning of these words:
   (a) labour
   (b) afterbirth
   (c) contraception

What foods should a pregnant woman eat?

Why do doctors recommend that pregnant women should not drink alcohol or smoke cigarettes?

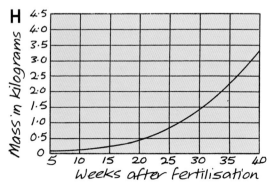

Growth of a foetus during pregnancy.

Children are usually free from disease during the first few months of their life. This is probably because a baby still has some sort of chemical protection that it received from its mother while it was attached to the placenta. These chemicals are called **antibodies** and they circulate inside the baby's blood, fighting microbes that could cause disease (**A**).

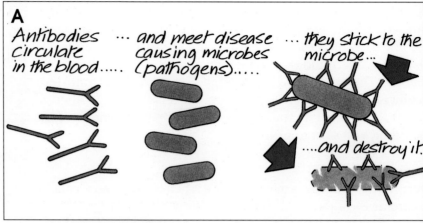

Antibodies fighting microbes.

For a long time, children were expected to catch several "childhood diseases" in the first few years of their lives, such as measles, German measles, chickenpox, smallpox, mumps, whooping cough, polio and diphtheria. It is clear the inherited protection against the microbes that cause these diseases, mostly bacteria and viruses (**B** and **C**), does not last long.

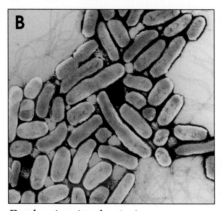

Food poisoning bacteria.

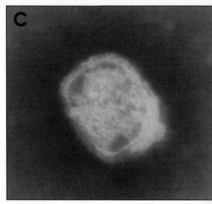

Smallpox virus.

Following the work of Edward Jenner and other medical scientists, over the last 200 years, we now understand more about how antibodies work. It seems that when the microbes get inside your body, special cells "recognise" that they shouldn't be there and they cause antibody chemicals to be made to destroy the "invaders".

These antibodies carry on circulating for several years and will kill the same microbes, if they appear again. Doctors can give you protection (or **immunise** you) against many childhood diseases by injecting you with dead or weakened disease-making microbes. Your body still makes the antibodies to protect you, but at very little risk of developing the disease.

D

Diphtheria, Whooping cough, Tetanus (triple) — 2-4 months

Polio (Given by mouth.)

Mumps, Measles, Rubella — 12-18 months

Rubella — Girls 10-14 years

Tuberculosis — 13 years

An immunisation timetable.

E

YEAR	EUROPE	AFRICA	S.E. ASIA
1920	180,000	2,000	100,000
1930	18,000	70,000	235,000
1940	4,000	15,000	200,000
1950	300	40,000	290,000
1960	50	16,000	40,000
1970	20	2,000	24,000
1980	0	0	0
1990	0	0	0

Cases of smallpox from 1920 until 1990.

Edward JENNER 1749-1823

JENNER'S CHIEF MEDICAL INTEREST WAS SMALLPOX. HE HAD HEARD AN OLD WIVE'S TALE THAT MILKMAIDS WHO CAUGHT COW-POX FROM COWS NEVER CAUGHT SMALL-POX. ALSO, THE TURKISH IDEA OF *'INNOCULATION'* WAS MUCH TALKED ABOUT AT THIS TIME. THIS INVOLVED INJECTING FLUID FROM BLISTERS OF PEOPLE WITH MILD CASES IN THE HOPE OF CATCHING A MILD CASE ONESELF.

JENNER SUSPECTED THAT THE COWPOX STORY MIGHT BE TRUE BUT IT HAD TO BE TESTED. IN 1796, HE FOUND A MILK-MAID WITH COWPOX, TOOK FLUID FROM A BLISTER ON HER HAND, AND INJECTED IT INTO AN EIGHT-YEAR-OLD BOY. THE BOY CAUGHT COWPOX. TWO MONTHS LATER, *JENNER* INOCULATED HIM WITH SMALLPOX. IT WAS AN INCREDIBLE RISK BUT, FORTUNATELY, *JENNER'S* THEORY WAS RIGHT AND THE BOY DID NOT CATCH THE DISEASE

DWARD JENNER WAS BORN IN BERKELEY, IN LOUCESTERSHIRE, THE SON OF A CLERGYMAN. HEN HE FINISHED SCHOOL, HE WAS APPRENTICED D A SURGEON, BUT HE WAS A GOOD NATURALIST ND WAS GIVEN THE JOB OF PREPARING AND RRANGING ZOOLOGICAL SPECIMENS COLLECTED Y *CAPTAIN COOK* ON HIS FIRST VOYAGE TO THE ACIFIC. HE WAS OFFERED A POST AS NATURALIST N THE SECOND VOYAGE, BUT TURNED IT DOWN.

HENRY CLINE, SURGEON OF ST THOMAS' HOSPITAL, SUCCESSFULLY TRIED THE NEW *'VACCINATION'* NAMED FROM THE LATIN WORD FOR 'COW') AND IT WAS QUICKLY ACCEPTED EVERYWHERE AFTER THAT. THE ROYAL FAMILY WERE ALL VACCINATED AND, IN 1802, PARLIAMENT VOTED *JENNER* £10,000 WHICH HE USED TO START A FOUNDATION TO SPONSOR VACCINATION. WITHIN 18 MONTHS, THE NUMBER OF SMALLPOX CASES IN ENGLAND WAS REDUCED BY TWO-THIRDS.

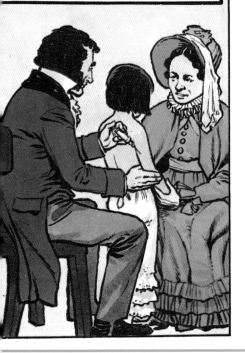

1. Name two childhood diseases and the sorts of microbe that cause them.

2. Why is it unlikely that a one-month-old baby will catch any of the usual childhood diseases?

3. Explain in your own words how antibodies fight disease.

4. What diseases have you been immunised against? You may need to ask someone at home to find out.

5. What hypothesis did Jenner test when he injected the eight-year-old boy with smallpox microbes?

6. Table **E** shows how cases of smallpox have changed from 1920 until 1990. Why do doctors believe that the fight against smallpox has been a success?

GROWING UP

The first five years or so of human life are a time of great change. A child's body grows very quickly and there is a lot of learning and exploring to do (**A**, **B**). It is a time when close bonds are made between the young child and those that care for him or her. Most people believe that a loving and secure early childhood sets a pattern for stable emotional growth into an adult.

Baby with mum and dad.

Toddler at play.

*How can you tell that this child is older than **B**?*

The bones of girls and boys grow at different rates. **D** shows the changes in appearance of children from 2 to 14 years old. The pictures are drawn to scale.

1. Here is a list of some stages in the development of a child over the first two years.
   A - can focus eyes on an object held at arm's length
   B - can say understandable words
   C - can reach out and touch an object
   D - can grip someone's finger strongly
   E - can draw a recognisable picture
   F - can call out to show they are hungry
   G - can take a few steps without help
   H - can hold a cup and feed themselves
   I - can attract attention
   Which stages are likely to happen at each of the following ages?
   (i) birth   (ii) 6 months   (iii) 12 months   (iv) 18 months
   (v) 24 months

2. What explanations can you give for these parents' comments?
   (a) "Ahmed is going through the 'terrible-twos'."
   (b) "I can't get Susie to play with other children."
   (c) "Sam often plays with a make-believe friend."

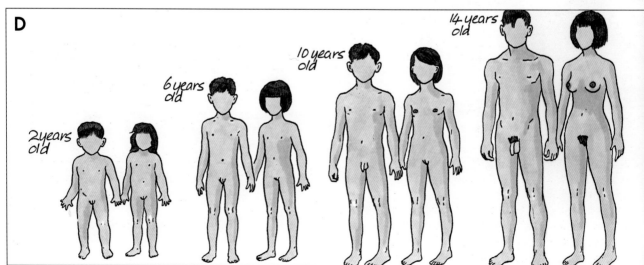
Children at 2, 6, 10 and 14 years.

3. Measure the heights of the boy and girl shown at the ages of 2, 6, 10 and 14 years. Work out their true heights using the scale (1cm on the page = 15cm in real life).

4. Use the measurements you have made together with the information given in table **E** to draw two line graphs of the growth of boys and girls up to 18 years.

E

Age (years)	Height (cm)	
	Boys	**Girls**
Birth	51	50
4	104	103
8	131	128
12	150	153
16	172	162
18	175	162

The physical changes we associate with "growing up" are matched by changes in our feelings that we cannot always explain. The chemicals called **hormones** that produce the changes in the bodies of children as they go through puberty, also cause changes in mood and emotions. The people around us, our family and friends as well as society in general, have certain expectations of how we should behave. It is not always easy to fit the picture we have of ourselves to the person that we appear to other people.

Becoming an "individual", with your own beliefs and attitudes, is a difficult process that often produces conflict (**F**, **G**).

Conflict is often associated with adolescence.

What might be the cause of this conflict?

5. A baby needs food, warmth and protection.
   (a) Are there any other things you would add to that list?
   (b) Make a list of the "needs" of people of your age.
   (c) Try to make a list of the needs of adults.
   (d) Discuss with a friend or in a wider class discussion to what extent "needs" depend on age, sex, or cultural background.

Now try this . . .

Graph **H** shows that some parts of the body grow at different rates.

Suggest reasons why the graph lines are different shapes.

H

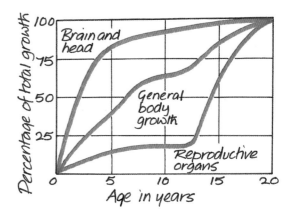

STAYING HEALTHY

When you are young you think you can eat as much or as little of anything that you like, do very little exercise and still live forever!

Part of the "growing up" process is to appreciate that the human body is not a robot that will never wear out - it is a living machine that needs constant maintenance (**A**). The fast pace of modern life means that we need to develop a lifestyle early in our lives that will keep us healthy for many years.

There are a lot of complex chemical reactions going on inside your body. If any of those reactions goes wrong, you may become ill. Many of the causes of illness (**B**) can be avoided if you lead a sensible lifestyle.

A

How healthy are these people?

B

Overweight people have more fat in their blood. This can affect the heart and blood vessels. Organs such as kidneys, liver, heart, become fatty and cannot work properly.

Fat deposits on the inside of blood vessels, makes them narrower. This can lead to 'heart attacks', 'strokes' and high blood pressure.

Excessive alcohol can damage brain, liver and kidneys.

Not eating the right foods causes health problems due to excess sugar, fat and lack of fibre.

Lack of exercise increases stress on heart and lungs caused by incorrect diet.

1 in 4 smokers suffer from diseases directly caused by smoking, such as heart disease, bronchitis and lung cancer.

How healthy is this man?

C

Things that can harm your health	Things you must do to stay healthy

1. Get into a group of four and draw a table like the one in **C**. Write down the five most important things that you should do to stay healthy. Write down five things that can harm your health.

2. Decide on a way of collecting the ideas of other groups and produce a class list for each column of the table. Discuss the reasons for any differences between the group lists.

A BALANCED DIET

Deciding which are the right foods is not an easy thing to do. We all need a certain amount of food to give us energy. There are three groups of energy foods - **proteins**, **fats** and **carbohydrates** (**D**).

D

A balanced diet. Do you have one?

It is dangerous to cut out any of those energy foods altogether. The best advice seems to be to reduce the amount of fat and **sugar** that we eat and meet more of our energy demands by eating **starchy** foods, such as potatoes, bread and rice (**E**).

The same plant foods, plus cereals and fresh fruit, contain a lot of what is called **dietary fibre** or **roughage**. This is the fibrous part of the plant that we cannot digest or use in our own bodies. It seems to be very important in helping food to pass through our intestine and may protect us from some sorts of cancer (**F**).

How can you improve your diet by taking their advice?

Food manufacturers are becoming more aware of the interest that people are taking in what they eat. Many of them now label their foods to show that they are low in salt, sugar and fat or high in vitamins and fibre (**G**).

How important is the labelling of food?

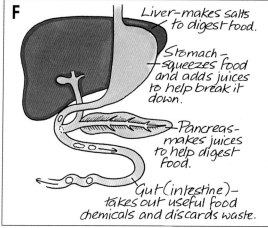

Your digestive system.

3. Make a display of food labels that represents what you think is a balanced diet. How close is it to your own diet?

4. Three students, Colleen, Kalid and Jemma, carried out a survey on 50 men and 50 women between 18 and 30 years old. They wanted to find out their views on health. They put their findings into a table (**H**).

   Discuss these comments made by each student:

   Colleen: "Our results show that men are heavier smokers than women."
   Kalid: "Most people know that smoking cigarettes can cause heart disease."
   Jemma: "Most people seem to know that exercise is good for your health."

H

Do you smoke cigarettes?	Yes = 26 females	Yes = 18 males
Can these be caused by smoking?	LUNG CANCER	HEART DISEASE
YES	95	62
NO	3	30
DON'T KNOW	2	8
Can these help to cause heart disease?	TOO MUCH FATTY FOOD	LACK OF EXERCISE
YES	82	23
NO	8	22
DON'T KNOW	10	55

THE ELECTRICITY MAKERS

The electricity that we "plug into" with our T.V.s, stereos, washing machines and so on is probably generated many miles away, possibly even in another country! Picture **A** shows a modern power station that generates electrical energy from burning coal.

Energy is released when a fuel is burned with the oxygen in the air. While some power stations burn coal, others use gas or oil, while a small number start with nuclear fuel. Whatever the fuel, the process of electricity generation involves using the heat from burning the fuel to boil water and change it into steam. The steam is very hot and at a high pressure, a high enough pressure to force huge turbines to turn (**B**).

What are the good and bad points of coal-fired power stations

The spinning turbines are connected to the wire coils of large electromagnets carrying an electric current. As you may remember from the Unit "Magnets and Spinning Coils" in Book Two, a coil spinning in a magnetic field generates an electrical voltage. This high voltage - 25,000V - is made even higher (up to 400,000V) by a device containing more wire coils, called a transformer. This huge and potentially dangerous electrical voltage is then distributed around the country in a network of electricity power lines called the **National Grid (C)**

Steam turbines generate electricity.

1. What is the purpose of the funnel-shaped towers that are built next to power stations?

2. Scientists have worked out that only about one-third of the energy stored in the fuel for the power station ends up as electrical energy in the National Grid. Suggest three "escape routes" for the remaining two-thirds of the energy.

3. Why are the cables that carry electricity so high above the ground?

4. Your home or school uses electricity supplied at 240V. Somewhere close to your home or school there will be a transformer that reduces the very high National Grid voltages to 240V. Find the transformer and describe what it looks like. What do the notices say that are attached to the transformer building?

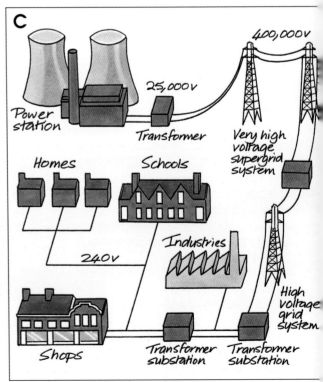

What is the function of the transformers?

Although power stations of the type described on page 80 produce most of the electricity we use, many people are concerned about the possible damage that the process of making electricity may be causing to our environment. Scientists are looking closely at different ways of generating electricity and are urging governments to think carefully about how energy will be provided in the future.

In a class discussion about the advantages and disadvantages of different sources of electrical energy, the students were asked to find out about the alternatives from wherever they could and to think about their own opinions.

D

We've got a lot of coal and it provides thousands of jobs.

Coal mining is dangerous and burning coal produces lots of smoke and dangerous gases.

Gas is a clean fuel and easy to transport through pipes.

Gas is expensive and explosive. It'll run out eventually.

Solar energy is the answer because it's environmentally friendly, but it only works well in sunny countries.

Oil is cheap, except when there's a war on.

Hydro-electricity doesn't cause pollution, but you need mountains and lots of water.

Nuclear fuel produces a lot of dangerous radioactive waste. Where will it go?

Nuclear energy is clean and efficient.

Oil makes too much smoke when you burn it. It makes carbon dioxide that adds to the greenhouse effect.

Do you agree with their opinions?

5. Explain what you understand by the "greenhouse effect".

6. (a) Where in the British Isles would you expect to find a nuclear power station?
   (b) What possible places can you suggest for the radioactive waste produced by a nuclear power station?

7. (a) What is hydro-electricity?
   (b) Where in the British Isles might you expect there to be hydro-electric power stations?

8. Summarise the information described by the students in a table of ADVANTAGES and DISADVANTAGES of the different energy sources. You may like to add your own opinions or those of others, discovered by using a survey.

Now try this . . .

Prepare for your own discussion of the alternative sources of electricity by researching the subject in a library or on an information data-base.

It is hard to imagine life for most of us without electricity. We take for granted that there will be light whenever we want it, for example. However, in the early part of this century, there was no national system for supplying electricity to homes, schools or workplaces (**A**).

1. Suggest what types of energy people used a hundred years ago for lighting, heating and cooking.

2. There are some parts of the United Kingdom that still do not have mains electricity.
   (a) Think of a reason why that might be.
   (b) How would your daily life be different if you lived in one of those places?

The more items of electrical equipment we have, the more we must pay for the electricity that they use. The kitchen shown in **B** has quite a few objects that use electricity.

What is the function of the circular objects attached to the wires on an electricity pylon?

Which of these appliances is necessary to a modern lifestyle?

3. Copy table **C** and list the electrical equipment you can find in the kitchen.

C

Electrical item	Electrical energy is changed into ...
Cooker	heat and light

4. Electrical energy is changed into other forms of energy by those objects. Next to the name of each object, say briefly what energy change is carried out. **For example:** a cooker changes electrical energy into heat and light.

Some of the devices are left on all the time, while others are only switched on when they are needed. If you look at your electricity meter at home, as different objects are switched on and off, you can see the meter numbers changing at different speeds (**D**). This shows that different devices use different amounts of electricity.

A modern electricity meter.

5. The graph lines in **E** show how much electricity is used by three devices, an electric clock, a refrigerator and an electric light.
(a) Which device goes with which graph?
(b) Sketch similar graphs for a washing machine and an automatic electric kettle.

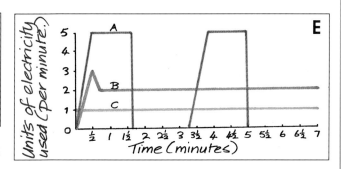

Each of the electrical devices has a plate on the back or underneath that shows how much electrical energy it uses. Two examples are given in **F** and **G**, but you should find some for yourself.

A kettle.

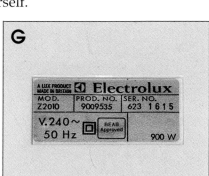

A vacuum cleaner.

Scientists measure the amount of electrical energy used by a device to carry out its job in units called **Joules**.

A device that uses 1 Joule of energy in every second is said to have a **power** of 1 **Watt**.
Example: an electric fire that uses 1 kilowatt means that it uses 1 kilojoule (1000 Joules) of energy in every second.

6. Copy table **H** and fill it in for the devices **F** and **G**, as well as any others you can find.

The electricity user must pay for the amount of electricity recorded on the meter. The electricity bill shown in **I** shows the number of electricity units used over the previous three months. Some of the units are called "night" units, while the rest are "day" units.

H

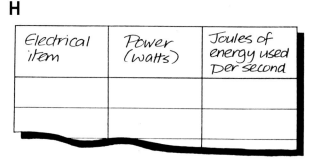

Electrical item	Power (watts)	Joules of energy used per second

7. (a) Why should the cost of electricity be cheaper if it is used at night?
(b) What sorts of electrical devices are likely to be working at night?

8. How can people change the way they use electricity at home to take advantage of the cheaper night units? Would it work in schools?

9. The electricity bill (**I**) was produced for a house in which electrical devices were switched on for a total of 420 hours. Work out the total power rating (in kilowatts) of devices.

Units of electricity = power x time
used (kW) (hrs)

An electric fire that uses 1 kilowatt in 1 hour has consumed 1 unit.

I

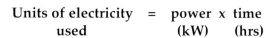

METER READINGS •		Tariff Code	UNITS	PENCE PER UNIT	AMOUNT EXCLUSIVE OF TAX	VAT REG No 567 9263 90	
Present	Previous					Tax	% Rate
02839	02534	10	305	2.380	7.26	NIGHT UNITS	
07625	07016	10	609	6.620	40.32	DAY UNITS	
STANDING CHARGE – SEE OVERLEAF					11.37		
			TOTAL ENERGY CHARGE		58.95	0.00	0.00

PLEASE QUOTE YOUR CUSTOMER REFERENCE NUMBER
IF YOU NEED TO CONTACT US REGARDING THIS BILL.

TARIFF CODE
10 DOMESTIC ECONOMY 7   58.95   0.00

PROGRAMMED READING DATE		THIS AMOUNT (incl. tax) IS NOW DUE
11 FEB 91	£ 58.95	Please pay before: 26 FEB 91

Electricity bill using "night" and "day" energy units.

ELECTRICITY IN CONTROL

We say that many of the electrical devices we use at home or at school are automatic. We usually mean that they turn themselves on or off according to what's happening around them. To do this they must be able to sense changes around them and react in some way.

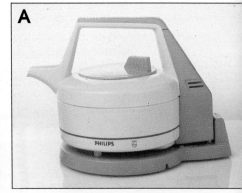

A

> Once switched on to boil water, the automatic kettle in **A** is able to "sense" a change which makes it turn off the supply of electricity to the heating element.
>
> **1.** Suggest what "change" in its surroundings the kettle responds to.

Why is it useful for a kettle to be automatic?

A car engine works best at a certain temperature. If it gets too hot, the moving parts may seize up and stop working altogether! When it is too cold, it is inefficient and does not burn the fuel properly.

The temperature of the engine is kept close to the best value by switching on and off the flow of cooling water and air inside and around it (**B**).

Modern cars have a special device called a temperature **sensor** which checks the temperature of the water circulating around the engine. As the working engine gets hotter, the sensor is able to switch on the cooling fan and open the valve that allows water surrounding the engine to circulate through the radiator and become cooled by the moving air.

Temperature sensors (**C**, **D**) are usually made from a piece of material which is a good conductor of electricity when hot, but a poor conductor when cold. Heating the material lowers its resistance to the movement of electrical current so that a device is switched on. A simple circuit component made from this material is called a **thermistor**.

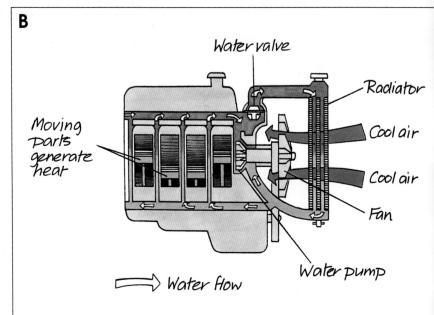

B

Water valve

Radiator

Cool air

Cool air

Fan

Moving parts generate heat

Water pump

⟹ Water flow

Why is it important to be able to sense the water temperature in a car engine?

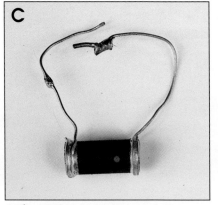

C

A thermistor.

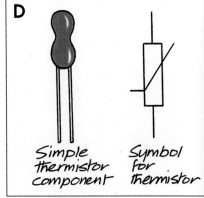

D

Simple thermistor component

Symbol for thermistor

Symbol for a thermistor.

> **2.** Suggest the names of two devices found in many homes that could make use of thermistors.

In this activity you will be able to investigate how temperature affects the size of an electrical current that can pass through a thermistor.

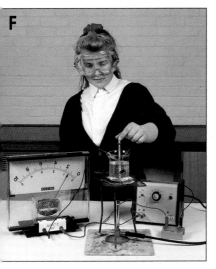

F

What you need

thermistor
glass beaker or other vessel suitable as a water bath
Bunsen burner, tripod and heat-proof mat
thermometer
milli-ammeter
safety resistor of 1,000 ohms
d.c. electrical supply (4.5 - 9V)
electrical leads and connections

 Wear goggles if you use a Bunsen burner to heat the water.

What you do

- Set up the circuit in **E**.
- Record the temperature of the water.
- When your teacher has checked your circuit, switch on the supply and measure the current flowing through the thermistor (**F**).
- Repeat your measurement of current at different temperatures up to 100°C. Keep electrical leads away from sources of heat.

Observations

Make up a table for your measurements of temperature and current.

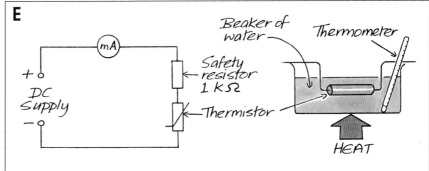

Circuit diagram for measuring current in a thermistor.

3. Draw a line graph of temperature against current. The temperature scale should be on the horizontal axis.

4. Describe the shape of the graph in words in order to say how the current flowing changes with temperature.

5. You could use your thermistor as a thermometer. How could your graph help you to do that?

6. From your observations, what would you say are the lowest and highest temperatures that your thermistor can measure accurately? Explain why you have chosen those numbers.

Now try this . . .

The electrical resistance of some materials changes when light is shone on them. **Light-dependent resistors** (**LDR**s) allow more current to flow in bright light than in dim light or in the dark (**G**).

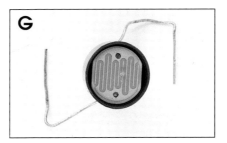

G

Design an investigation to find out how an LDR could be used to measure the intensity of light produced by different makes of light bulb.

MAKING DECISIONS

The "on" and "off" switching used in electrical devices shows you that even complicated-looking machines work by simple rules - even if they do it very quickly! The system of "rules" that controls what decisions are made inside electronic circuits is usually called **logic**. Inside computers, for example, there are huge numbers of tiny switches collected together into "**chips**" (**A**). At any time, some of the switches will be "on" while others are "off". The pattern of "on"s and "off"s decides how the computer reacts to messages put into it.

You will be able to understand the rules of logic by looking at some simple circuits.

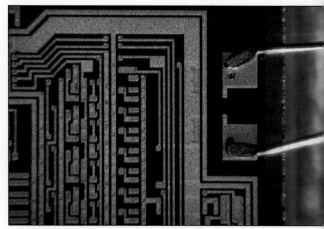

What is the real distance between the connections to this chip? (Magnification = 100x)

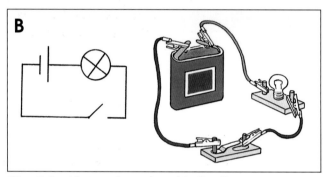

Circuit 1 with switch open and bulb off.

RULE 1: When a switch is open, electricity cannot flow and a bulb will not light up. The state of the switch can be written as "0".

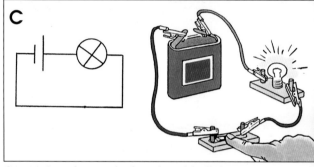

Circuit 2 with switch closed and bulb lit.

RULE 2: When a switch is closed, electricity flows and a bulb will light up. The state of the switch can be written as "1".

Scientists describe the state of the switches in a circuit and the effect on the bulb, or whatever the circuit operates, using tables like **D** and **E**. These are called **truth tables**.

D

Circuit	State of switch	State of bulb
1	open	off
2	closed	on

Truth table for circuits 1 and 2 using words.

E

Circuit	State of switch	State of bulb
1	0	0
2	1	1

Truth table for circuits 1 and 2 using symbols.

1. Look at the circuits 3 - 6 (**F**). Predict if the bulbs will light in each case.

2. Copy the truth table (**G**) for those four circuits and complete it.

F

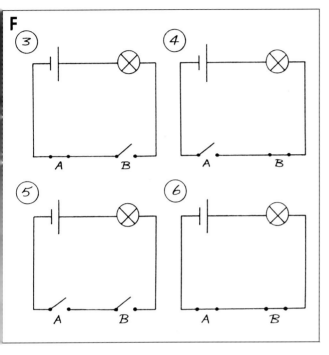

The circuits 3 - 6 are all called "**AND**" circuits. That is because the bulb will only light if both switch 1 AND switch 2 are closed (both "1" in the truth table).

Modern electronic devices do not have switches that look like the type you are familiar with when you build circuits. Electronic switches are themselves small circuits called **gates**. Gates, as you would expect, either allow electricity to pass or they do not, depending on the type of gate.

H is the symbol for a gate that would work in the same way as circuits 3 - 6. It is an **AND** gate.

H

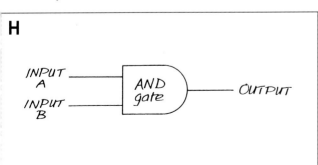

AND gate with 2 inputs.

G

Circuit	State of switches		State of bulb
	A	B	
3	1		
4	0		
5		0	
6		1	

There are several other types of logic gate. **I** shows the symbol for an **OR** gate together with the **OR** circuit that it represents. In an OR circuit, a closed switch at either **A** OR **B** will light the bulb.

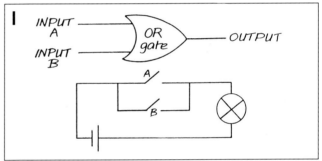

OR gate symbol and circuit.

J

State of inputs		State of output
A	B	
0	0	
0		1
	0	1
1	1	

Truth table for OR gate.

3. Copy and complete the truth table (**J**) for the OR gate circuit shown in **I**.

4. Suppose that you have been given a gate circuit sealed into a box so that you cannot see into it. What could you do to find out if the gate is an AND or an OR gate circuit?

INFORMATION HANDLERS

In previous units you have read about sensors or **input devices** that collect information from the surroundings and pass it on to some sort of decision unit or **logic gate**. You have learned that the logic gates handle the messages according to simple rules before sending signals to lamps, motors or another **output device**, which performs some sort of response.

Diagram **A** summarises those ideas. It shows how similar electronic information handling is to the way we react to, say, putting our hand on a pin. That sort of "information handling" is called a reflex (**B**).

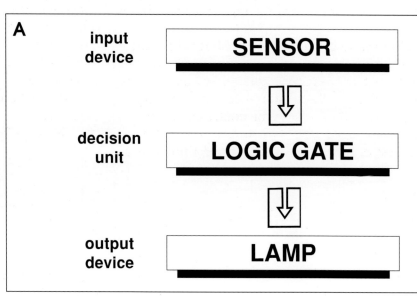

A simple electronic system.

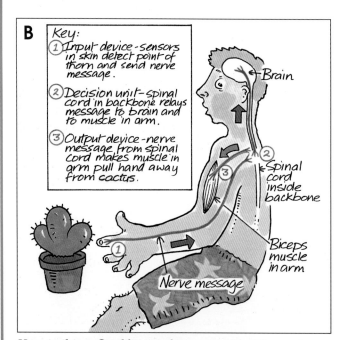

How is this reflex like an electronic system?

Many devices, including washing machines, microwave ovens and modern cameras, take this two-step process further and relay the message to a memory unit that stores the information. The memory may also have information programmed into it that can be compared with the signals coming in from a set of logic gates or processor. This allows the device to check constantly what it is doing and react accordingly (**C**).

1. Which of the following items are input devices and which are output devices?
   - light bulb
   - motor
   - thermistor
   - loudspeaker
   - bar code reader
   - microphone
   - buzzer
   - light-dependent resistor (LDR)

2. In the example of the human reflex, say what parts of the body represent the input device, the processor and the output device.

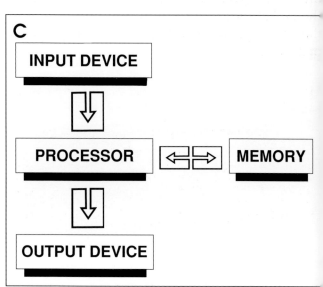

An electronic system with memory.

Electronic systems have to be able to handle two sorts of information, called analogue and digital.

Analogue information is the sort of signal produced by sensors and most input devices. "Analogue" means that it varies continuously, such as the signal produced by a thermistor recording changes in temperature (**D**). A volume control on a radio is also an analogue device because it can change the loudness of a sound in a smooth, continuous way.

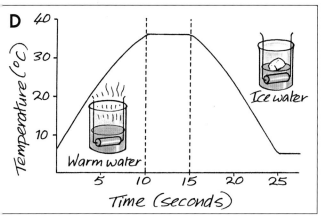

Graph showing the output from an analogue thermistor.

Digital information is the only sort of signal that can be processed directly by a processor or an electronic memory. "Digital" means that the signal goes up or down in jumps, without any value in between. Switches that are "on" or "off" are digital devices, so logic gates that work like switches also process information digitally. The sound information recorded on a stereo compact disc is of the digital sort, where the coded information tells the disc player when to make a sound and when to make no sound (**E**).

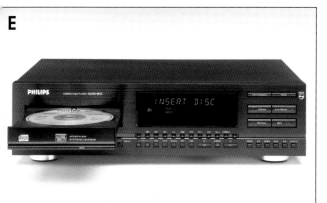

Compact discs - a modern type of digital information.

ANALOGUE OR DIGITAL?

Bar codes are now used on many products in shops (**F**). The items carrying a bar code are quickly passed over a bar reading machine which shines a fine laser light beam across the strip of bars and spaces. The white areas reflect the beam back to a sensor while the black bars absorb the light (**G**).

Why are bar codes useful on supermarket shelves?

Laser beam reading a bar code.

This information is passed back to a processor which compares the signal with information about the products stored in its memory. A bill is printed for the customer and the stock numbers are updated.

3. Is a bar code an example of analogue or digital information?

4. Design a simple bar-and-space code so that you can recognise a five-letter word. You can vary the number and thicknesses of the lines.

5. Suggest another example of a device that can be used to identify a person electronically.

Now try this . . .

Some televisions can be turned on and off using remote control handsets. Suppose you wanted to turn the T.V. on from another room using the remote control.

Can the instructions from the remote control be bounced off walls, doors or other surfaces to reach the T.V.? If you have access to a device with a hand control, try sending instructions in this way.

SENDING MESSAGES

Electric current has been used to carry messages since early in the 19th century. The electric **telegraph** made use of a code specially invented for it by Samuel Morse during the 1830s (**A**). His system involved sending short and long pulses of electricity down a wire which could be "translated" or decoded either as sounds by a buzzer or as dots and dashes by a printer (**B**).

A

Samuel Morse.

B

A	•‑	J	•‑‑‑	S	•••
B	‑•••	K	‑•‑	T	‑
C	‑•‑•	L	•‑••	U	••‑
D	‑••	M	‑‑	V	•••‑
E	•	N	‑•	W	•‑‑
F	••‑	O	‑‑‑	X	‑••‑
G	‑‑•	P	•‑‑•	Y	‑•‑‑
H	••••	Q	‑‑•‑	Z	‑‑••
I	••	R	•‑•		

The Morse code alphabet.

What you need

4.5V battery or other d.c. supply
connecting leads and clips
buzzer and/or light bulb (3V or 4.5V)
switch

What you do

● Using the equipment provided, put together a circuit which will allow you to send short or long light flashes or sounds (**C**).
● Use Morse's code (dot = short, dash = long) or make up a simple code for yourself.
● Send a coded message to your partner.

Can you send the message round a corner?
What are the limitations to a code such as this?

C

Sending messages by light.

Then in 1876 a Scotsman, Alexander Graham Bell, invented the first reliable device for sending speech over long distances, the **telephone**. Bell had to develop two parts to the telephone, a **microphone** and an **earpiece**.

A microphone changes sound waves into a varying electric current (**D**) while the earpiece converts the changing current back into sound waves. The pattern of the electric current must be a close copy of the sound waves made by the person speaking.

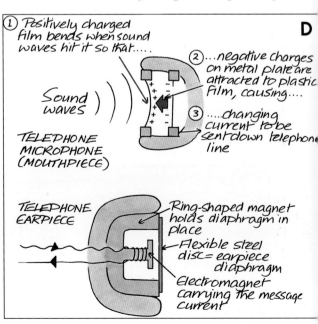

D

① Positively charged film bends when sound waves hit it so that.....

②...negative charges on metal plate are attracted to plastic film, causing....

③ .....changing current to be sent down telephone line

Sound waves

TELEPHONE MICROPHONE (MOUTHPIECE)

TELEPHONE EARPIECE

Ring-shaped magnet holds diaphragm in place

Flexible steel disc = earpiece diaphragm

Electromagnet carrying the message current

Structure of telephone microphone and earpiece.

CARRYING THE MESSAGE

An important feature of the telephone system is the ability to be connected to a large number of people. This means that some sort of switching is needed to direct a call to the right destination. At first, **exchange** operators (**E**) made a connection between the caller's telephone wires and those of the person receiving the call!

An operator connecting a call.

Of course, the number of calls now being made means that a much faster, automatic system is required. Modern telephone exchanges change the analogue voice signals into digital information that electronic switches and computers can control (**F**). These automated exchanges are able to combine electronically about one thousand signals and transmit them along a single wire, to be "unscrambled" by other local exchanges and relayed to a particular subscriber.

In the 1960s scientists realised that there could be a better and cheaper way of carrying messages than as electric current in a copper wire. It was known that if a ray of light was shone into one end of a glass fibre of a certain purity and size, it would be reflected from side to side until it appeared at the other end nearly as bright as at the beginning!

A modern electronic exchange.

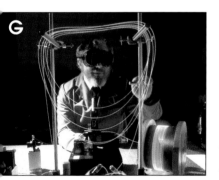

Testing optical fibres.

With the help of the right sort of sensors, scientists found a way of changing variations in a speaker's voice into changes in the light rays sent down the glass **optical fibre**. They also discovered a way of turning the pulses of light that emerged at the other end of the fibre back into sound. Modern optical fibre communication now uses a pure form of light - **laser** light - and can carry over 10,000 messages at once!

H shows a piece of office equipment which has added a new dimension to the use of the telephone network. It is a facsimile (fax) machine.

How is a fax message different from the usual sort of message sent by telephone?

In what situations is it more useful to be able to send information by fax?

1. Explain the difference between the functions of a microphone and a loudspeaker.

2. Name two advantages of having electronic exchanges instead of operators to handle telephone calls. Are there any disadvantages?

3. Imagine you wanted to lay a telephone cable across the Atlantic Ocean, between the United States and the United Kingdom. Would you choose copper wire cables or optical fibres? Give reasons for your choice.

HERTZ
1857·1894

MARCONI
1874·1937

HEINRICH HERTZ STUDIED PHYSICS UNDER THE FAMOUS PHYSICISTS *HELMHOLTZ AND KIRCHHOFF.* IN 1885, THE BERLIN ACADEMY OF SCIENCE OFFERED A PRIZE FOR THE SOLUTION TO THE PROBLEM IN ELECTROMAGNETISM. *HERTZ* WAS RELUCTANTLY PERSUADED TO TRY. BUT IN 1887 HE DISCOVERED SOMETHING FAR MORE IMPORTANT ~ RADIO WAVES.

HERTZ SET UP TWO METAL BALLS, SEPARATED BY AN AIR GAP, WITH AN ELECTRICAL CURRENT OSCILLATING BETWEEN THEM. THE CURRENT JUMPED THE GAP, CAUSING AN OSCILLATING SPARK. ACCORDING TO EQUATIONS ESTABLISHED BY *JAMES CLERK MAXWELL,* THERE SHOULD BE ELECTROMAGNETIC RADIATION WITH SUCH A SPARK.

TO DETECT THE RADIATION, *HERTZ* USED A LOOP OF WIRE WITH A GAP AT ONE POINT. RADIATION CAUSED CURRENT TO FLOW IN THE LOOP, DETECTABLE BY SPARKS CROSSING THE GAP. BY MOVING THE LOOP AROUND HIS ROOM, *HERTZ* COULD CALCULATE THE SHAPE OF THE WAVES. HE FOUND WAVELENGTH WAS A MILLION TIMES THAT OF LIGHT.

HERTZ DIED AT THE EARLY AGE OF 37, AND SO DID NOT LIVE TO SEE *MARCONI* DEVISE A PRACTICAL USE FOR THE RADIATION HE HAD DISCOVERED. *GUGLIELMO MARCONI* WAS 20 WHEN HE CAME ACROSS AN ARTICLE ON *HERTZ'S* ELECTROMAGNETIC WAVES. HOPING TO USE THESE WAVES FOR SIGNALLING, HE BEGAN EXPERIMENTING ON HIS FATHER'S ESTATE.

HE USED *HERTZ'S* METHOD OF PRODUCING THE WAVES AT THE SIGNALLING END, CONTROLLING THE SPARK BY A *MORSE* KEY. AT THE RECEIVING END, HE USED A CONTAINER OF LOOSELY~PACKED METAL FILINGS THAT CONDUCTED A DETECTABLE CURRENT ONLY WHEN RADIO WAVES FELL UPON IT.

IN 1895, HE FIRST SENT A SIGNAL FROM HIS HOUSE TO HIS GARDEN. HE IMPROVED HIS RANGE TO 1½ MILES BY CONNECTING THE TRANSMITTER AND RECEIVER TO THE EARTH, AND USING A WIRE AS AN AERIAL.

MARCONI WAS GETTING LITTLE ENCOURAGEMENT IN ITALY, SO IN 1896 HE TRAVELLED TO ENGLAND, WHERE HE FILED HIS PATENTS. HE BEGAN A SERIES OF SUCCESSFUL EXPERIMENTS, SENDING SIGNALS 4 MILES ON SALISBURY PLAIN AND 9 MILES ACROSS THE BRISTOL CHANNEL.

MARCONI'S COUSIN, JAMESON DAVIS, HELPED HIM SET UP A WIRELESS TELEGRAPH COMPANY. THE FIRST COMMERCIAL MESSAGE WAS SENT BY THE VERY ELDERLY LORD KELVIN TO THE EVEN OLDER PHYSICIST GEORGE STOKES. IN 1899, MARCONI EQUIPPED 2 SHIPS TO REPORT TO NEWSPAPERS IN NEW YORK CITY THE PROGRESS OF THE YACHT RACE FOR THE AMERICA'S CUP.

IN 1897 MARCONI RETURNED TO ITALY. A LAND STATION WAS SET UP AT LA SPEZIA AND COMMUNICATIONS ESTABLISHED WITH ITALIAN WARSHIPS AS MUCH AS 12 MILES OUT TO SEA. THE NEXT YEAR HE ESTABLISHED A RADIO LINK BETWEEN SOUTH FORELAND LIGHTHOUSE AND THE EAST GOODWIN LIGHT SHIP IN BRITAIN.

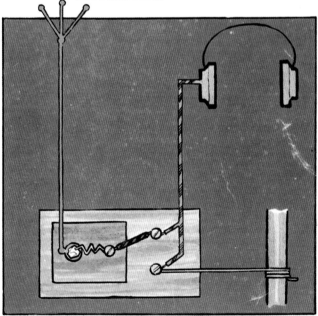

IN 1901 MARCONI SUCCEEDED IN SHOWING THAT RADIO WAVES FOLLOW THE CURVE OF THE EARTH. HE SENT A SIGNAL FROM CORNWALL TO NEWFOUNDLAND, USING KITES TO LIFT THE AERIALS AS HIGH AS POSSIBLE.

MARCONI CONTINUED WORKING TO IMPROVE RADIO COMMUNICATIONS ALL HIS LIFE. IN 1916, HE BEGAN WORKING ON SHORTER WAVELENGTH TRANSMISSIONS. IN 1924, THE MARCONI COMPANY OBTAINED A CONTRACT FROM THE POST OFFICE TO SET UP SHORT-WAVE COMMUNICATIONS BETWEEN BRITAIN AND AUSTRALIA. MARCONI WAS AWARDED MANY HONOURS FOR HIS WORK. IN 1909, HE SHARED THE NOBEL PRIZE IN PHYSICS WITH KARL BRAUN, WHO HELPED IMPROVE RADIO TECHNOLOGY. YOU CAN MAKE A RADIO RECEIVER FOR YOURSELF. YOU NEED A PIECE OF WOOD ABOUT 2X3X½ INCHES FOR A BASE·3 LARGE SCREWS·A PIECE OF BRASS CURTAIN ROD HALF AN INCH IN DIAMETER AND AN INCH LONG·A CRYSTAL OF IRON PYRITES OR GALENA·A PIECE OF PHOSPHOR-BRONZE WIRE FOR THE CAT'S WHISKER, ABOUT 3½ INCHES LONG·A PAIR OF EARPHONES·SEVERAL FEET OF INSULATED WIRE. SOME OF THE PIECES NEED TO BE SOLDERED, GET SOMEONE TO HELP YOU. DRILL A HOLE NEAR ONE END OF THE BASE, HALF AN INCH IN DIAMETER. HEAT THE END OF THE BRASS ROD IN A GAS FLAME AND RUN IT FULL OF MELTED SOLDER. EMBED THE CRYSTAL IN THE SOLDER AND FORCE THE ROD INTO THE HOLE IN THE BASE. SCREW ONE OF THE SCREWS IN NEAR THE OTHER END. TWIST ONE END OF THE PHOSPHOR-BRONZE WIRE ROUND THE SCREW. SHARPEN THE OTHER END, AND BEND AND ADJUST THE WIRE SO THAT THIS END LIGHTLY PRESSES ON THE CRYSTAL. MOUNT THE RECEIVER ON A PIECE OF BOARD AND NEXT TO IT SCREW IN THE OTHER SCREWS. WITH INSULATED WIRE, CONNECT THE SCREW ON THE RECEIVER TO ONE SCREW ON THE BOARD. CONNECT THE EARPHONES TO THIS SCREW AND THE OTHER ONE. THE RECEIVER NEEDS AN AERIAL TO PICK UP THE RADIO WAVES, AND IT NEEDS TO BE GROUNDED. USE AS LONG A PIECE OF WIRE AS POSSIBLE FOR AN AERIAL, CONNECTED TO THE BRASS ROD HOLDING THE CRYSTAL. YOU CAN GROUND THE RECEIVER WITH A WIRE ATTACHED TO ONE SCREW AS SHOWN AND WOUND ROUND AN IRON WATER OR GAS PIPE AT THE OTHER END. THIS RECEIVER IS NOT SELECTIVE ENOUGH TO PICK UP BROADCAST PROGRAMMES BUT YOU MAY BE ABLE TO PICK UP ELECTRICAL DISTURBANCES SUCH AS A PIECE OF MACHINERY BEING SWITCHED ON OR OFF. PUT ON THE EARPHONES AND MOVE THE POINTED END OF THE PHOSPHOR-BRONZE WIRE ABOUT ON THE CRYSTAL UNTIL YOU CAN HEAR SOMETHING. IF YOU HAVE MADE THE HENRY TELEGRAPH SENDER, YOU COULD PICK UP SIGNALS FROM THAT.

INDEX

ACKNOWLEDGEMENTS

The material in this, and the other books in the series owes much to the schools and colleagues each author has been working with over the years as often ideas have been progressively developed, distilled from the author's mind, modified from their privately developed worksheets or developed from original ideas.

The additional creators of work are here acknowledged, namely Maureen Barwick, Janet Bingham, Wendy Butcher, Brenda Jones, Pat Woodman and Simon de Pinna.

The authors and publishers gratefully acknowledge the assistance given by Mr. S. Peacey, Mrs. V. Jennings and pupils of Lea Manor High School, Luton, Bedfordshire in the production of the classroom photographs.

The work of Janet Tickle and of the artist, Pat Murray, is greatly appreciated.

The publishers have made every effort to contact copyright holders but this has not always been possible. If any have been overlooked we will be pleased to make any necessary arrangements.

Cover design: Abacus Publicity Limited
4a-f: Fisons plc
6a: Directorate of Public Affairs
12b: AEA Technology
13c-f: Simon & Schuster Young Books
14/15: Simon & Schuster Young Books
17c: Central Electricity Generating Board
　d: AEA Technology
　e: James Davis Photography
　f: Berthold (UK) Ltd
18a: K. Roberts/Science Photo Library
　b: Avon Cosmetics Ltd
　c: James Davis Photography
　d: Design Museum
　e: James Davis Photography
　f: British Gas
22a: National Medical Slide Bank
24d: Maggie Martin/Nottingham Polytechnic
26a: Solid Fuel Advisory Service
28b: Science Museum
　c: Science Museum
30a,b,c: Maggie Martin/Nottingham Polytechnic
31g: Imperial Chemical Industries
34/35: Simon & Schuster Young Books
36a: NASA/Starland Picture Library
　b: British Gas
37e: Lever Brothers Ltd
38b: Roberts and Belk Ltd
39d: Maggie Martin/Nottingham Polytechnic
40a: Tesco Creative Services
42a: NASA/Science Photo Library
43f: James Davis Photography
　g: John Warden/Natural Science Photos
　h: James Davis Photography
44a: James Davis Photography
　b: NASA/Starland Picture Library
　c: NASA/Science Photo Library
45f: National Optical Astronomy Observatories/Starland Picture Library
　g: Yerkes Observatory/Starland Picture Library
　h: Derek Aspinall/Starland Picture Library
　i: Yerkes Observatory/Starland Picture Library
48a: National Optical Astronomy Observatories/Starland Picture Library
　b: NASA/Starland Picture Library
　c: James Davis Photography
50/51: Simon & Schuster Young Books

52a: Secchi-Lecaque/Roussel-UCLAF/CNRI/Science Photo Library
　b: Jeremy Burgess/Science Photo Library
　c: CNRI/Science Photo Library
　d: James Bell/Science Photo Library
53i: Alastair MacEwen/Oxford Scientific Films
54b: Tony Rogers/Polytechnic South West, Exmouth
　c: Jeremy Burgess/Science Photo Library
56a: Stephen Dalton/NHPA
　b: Primrose Peacock, Holt Studios Ltd
　c: E. A. James, NHPA
58a,b: Harry Smith Horticultural Photographic Collection
62a: London Scientific Films/Oxford Scientific Films
64/65: Simon & Schuster Young Books
68b,c: Dr. A. Bush/Royal Brompton National Heart & Lung Hospitals
69e: Eric Graves/Science Photo Library
　f: London Scientific Films/Oxford Scientific Films
　g: Sickle Cell Society
70a: Eric Graves/Science Photo Library
　c: David Thompson/Oxford Scientific Films
　d: Ben Osborne/Oxford Scientific Films
71e,f: B. P. Kent Animals Animals/Oxford Scientific Films
　g: D. Yendall/Natural Science Photos
　h: I. West/Natural Science Photos
　i: P. Parks/Oxford Scientific Films
　j: J. Cooke/Oxford Scientific Films
73e: L. M. Beidler/Science Photo Library
74b: A. Dowsett/Science Photo Library
　c: London School of Hygiene and Tropical Medicine/Science Photo Library
75: Simon & Schuster Young Books
76a,b: James Davis Photography
77g: Sally and Richard Greenhill
78a: James Davis Photography
79g: Tesco Creative Services
80a,b: The Electricity Association
82a,d: The Electricity Association
84a: Philips Domestic Appliances
86a: John Walsh/Science Photo Library
89e: Philips CED Publicity
　f: Tesco Creative Services
　g: Paul Shambroom/Science Photo Library
90a: Institution of Electrical Engineers
91e: British Telecom Museum
　f,g: TeleFocus, a British Telecom photograph
　h: Ricoh
92/93: Simon & Schuster Young Books